Beaver Protection, Management, and Utilization in Europe and North America

Beaver Protection, Management, and Utilization in Europe and North America

Edited by

Peter E. Busher
Boston University
Boston, Massachusetts

and

Ryszard M. Dzięciołowski
Agricultural University SGGW
Warsaw, Poland

Kluwer Academic / Plenum Publishers
New York, Boston, Dordrecht, London, Moscow

Library of Congress Cataloging-in-Publication Data

Beaver protection, management, and utilization in Europe and North America / edited by Peter E. Busher and Ryszard M. Dzięciołowski.
p. cm.
"Proceedings of a symposium ... held as part of the Euro-American Mammal Congress, July 19-24, 1998, in Santiago de Compostela, Spain"--T.p. verso.
Includes bibliographical references and index.
ISBN 0-306-46121-8
1. Beavers--Ecology--North America Congresses. 2. Wildlife conservation--North America Congresses. 3. Beavers--Ecology--Europe Congresses. 4. Wildlife conservation--Europe Congresses.
I. Busher, Peter E. II. Dzięciolowski, Ryszard. III. Euro-American Mammal Congress (1998 : Santiago de Compostela, Spain)
QL737.R632.B425 1999
333.95'937--dc21 99-30049
CIP

Proceedings of a symposium on Beaver Protection, Management, and Utilization in Europe and North America, held as a part of the Euro-American Mammal Congress, July 19–24, 1998, in Santiago De Compostela, Spain

ISBN 0-306-46121-8

233 Spring Street, New York, N.Y. 10013

10 9 8 7 6 5 4 3 2 1

A C.I.P. record for this book is available from the Library of Congress

Printed in the United States of America

PREFACE

By the end of the 19th century both beaver species had been extirpated from large portions of their native ranges. The global decline in beaver populations was the direct result of exploitation by humans. Now, at the end of the 20th century, protection, management, and reintroduction programs, coupled with a decline in the demand for beaver fur and other products, have allowed beaver populations to increase dramatically. Since beavers actively modify their local environment their activities can conflict with human land use. Because of this, the beaver once considered a unique and exotic component of wetlands, is now often considered a nuisance species. The history, as well as the current status, of beaver populations in Europe and North America provide insight into how conservation programs work, and into how humans and wildlife interact.

The initial plenary lecture of the Euro-American Mammal Congress (July, 1998) was presented by Dr. Michael L. Rosenzweig, a professor at the University of Arizona. Dr. Rosenzweig discussed how humans have used and continue to use natural resources, including wildlife and wildland. He provided evidence indicating that the current model of reservation conservation could not provide a long-term solution to the human-wildlife/wildland conflict. Dr. Rosenzweig emphasized that what is required is a move away from purely exploitive activities (I would call this exploitive ecology) and the development of a reconciliation ecology with wildlife. The concept of reconciliation ecology — the development of new habitats that allow for both human use and self-supporting populations of other species — is the essence of beaver ecology. The concept of reconciliation ecology permeated the talks at the beaver symposium, it permeated informal discussions among the beaver researchers at the Congress, and it permeates the papers in this volume.

The two extant beaver species, the European or Eurasian beaver (*Castor fiber*) and the North American or Canadian beaver (*Castor canadensis*) practice natural reconciliation ecology. Beavers, more than any other species (excluding humans), influence their environments. Their activities shape the landscape, influence the biogeochemical cycles, and affect the biological diversity in their aquatic habitats. Because of this they are viewed as a "Keystone" species. However, they should also be viewed as a "Reconciliation" species since they provide a model for this approach to ecosystem management. Beavers, by their activities, develop new habitats that support continued beaver use, and also allow for self-supporting populations of other species. Beaver ecology is reconciliation ecology. Perhaps, by studying beaver ecology, humans can learn how to better co-exist with wildlife.

Beavers can also be used as a model to help understand human-wildlife relations. Historically, in North American and Europe, beavers were an important part of human culture. Beavers provided fur and other products that were economically valuable. This led to a decline in the population size and in many cases (especially in Europe) the complete extirpation of the species. Conservation efforts, especially in the past 50 years, have allowed beaver populations to increase, and have allowed beavers to recolonize many areas. Presently, increasing beaver populations and humans are often in conflict over land use. While the human population continually alters and exploits the landscape, especially wetlands, beavers, as natural reconciliation ecologists, attempt to reclaim and develop the land in a different manner. The reclamation activity of beavers produces a habitat that is more suitable for wildlife. Thus, beaver populations which were reduced by human activities, then "saved" by conservation efforts, are now potential competitors with humans for land and other resources. What is necessary is a strategy of reconciliation ecology for beavers and humans.

This volume is the result of a symposium, *Beaver Protection, Management and Utilization in Europe and North America,* that was part of the Euro-American Mammal Congress (EAMC) held in Santiago de Compostela, Spain, 19–24 July, 1998. The editors of this volume also served as the conveners of the symposium. The papers in this volume provide information on the current status of beavers in many regions of Europe where only a few decades ago they did not exist. The papers on reintroduction efforts in The Netherlands, Austria, the Baltic States, Poland and Russia all illustrate the problems and prospects facing beavers in the human-dominated landscape of modern Europe. The interaction between large-scale, human development, such as dams and hydroelectric plants, and beavers is documented in the papers on the Slovakian beaver population. The European papers also provide analyses of different management models and concerns. The North American papers provide information on long-term population dynamics of unexploited populations, and examine how beaver behavior may be useful as a management tool. This volume is unique in that it includes reports from both Europe and North America on beaver ecology, beaver management programs, beaver conservation efforts, and on the continuing effort to reintroduce this "natural ecologist" into its native range. The essence of the beaver symposium and the papers in this volume is a reconciliation between humans and beavers.

ACKNOWLEDGMENTS

The editors, who acted as the co-conveners of the symposium, are grateful to all of the scientists who participated in the symposium. We would also like to acknowledge the support supplied by the organizing committee of the Euro-American Mammal Congress, the Universidad de Santiago de Compostela, especially the Facultades de Farmacia y Bioloxia, the Sociedad Española para la Conservación y el Estudio de Mamíferos (SECEM), the Societas Europaea Mammalogica (SEM), the American Society of Mammalogists, and Xunta De Galicia (the autonomous government of the Galician Region). We also acknowledge the support of our institutions, the College of General Studies, Boston University (PEB), and the Department of Forest Zoology and Wildlife Management, Agricultural University SGGW, (RMD). Editorial support was provided by Patricia Mullane (Div. Science and Mathematics, CGS, Boston University).

CONTENTS

BEAVER MANAGEMENT AND UTILIZATION IN SCANDINAVIA

Göran Hartman

Department of Conservation Biology
Swedish University of Agricultural Sciences
Box 7002, 750 07 Uppsala, Sweden

ABSTRACT

The European beaver (*Castor fiber*) was once abundant throughout Scandinavia. Due to over-hunting beavers were exterminated in Sweden and Finland during the second half of the 19th century. Because of a prudent ban on beaver hunting in 1845, a small remnant population of a few hundred individuals survived in southern Norway. This population became the source for subsequent reintroductions to the other Scandinavian countries. Between 1922 and 1940 about 80 Norwegian beavers were reintroduced to Sweden, and in 1935–36, 17 animals were released in Finland. In 1937, seven North American beavers (*Castor canadensis*) were also introduced to Finland. The reintroductions were successful and current population sizes are estimated to be 100,000 in Sweden, 50,000 in Norway, and 10,000 in Finland. However, about 90% of the Finnish population are North American beavers. The populations are still increasing, although the increase is slow in Finland. Hunting regulations differ substantially between the three countries. Sweden has the least elaborate system (for example, no bag limit), and Norway the most sophisticated. It is obvious that there are other factors affecting hunting intensity since the annual beaver harvest is only 6% of the population in Sweden, 10% in Norway, but as much as 20% in Finland. Many landowners consider beavers a nuisance, and therefore a future task is to increase public awareness of the value of beavers. Beavers represent not only a game species, but are also important ecological engineers. Beaver activity increases biological diversity and influences the hydrologic regime in a landscape dominated by commercial forestry.

1. INTRODUCTION

Like in most parts of the world, the Scandinavian beaver populations were hunted to or near extinction. All that was left at the turn of the 20th century was a small population

Beaver Protection, Managment, and Utilization in Europe and North America, edited by Busher and Dzięciołowski.
Kluwer Academic / Plenum Publishers, New York, 1999.

of a few hundred individuals in the southern parts of Norway. During the 1920s and 1930s this became the source for reintroductions to Sweden and Finland. Unfortunately a few beavers of the North American species (*Castor canadensis*) were also introduced to Finland. In Denmark, beavers had been hunted to extinction during the bronze age (Aaris-Sørensen, 1988). The reintroductions and following translocations were successful. The beaver has reclaimed large parts of its former range (Figure 1), and reached substantial population sizes.

We estimate there to be about 100,000 beavers in Sweden (Hartman, 1995), 50,000 in Norway (Nolet and Rosell, 1998) and 10,000 in Finland (Lahti, 1997). However, it is unfortunate from a conservation point of view, that about 90% of the Finnish population is the North American species. The ranges of the two species do not overlap in Finland. The European beaver is mainly found in an area between the Baltic coast in the west, and lake Nasijarvi in the east. Denmark has not yet tried to reintroduce beaver, but a feasibility study by Ina Fischer-Andersen (1997) has shown that they probably could harbor a small population. It has not been decided if there will be an attempt to reintroduce beaver to Denmark.

The population development during the last four decades is presented in Figure 2. The increase has been rather dramatic in Norway and Sweden but not in Finland, most likely due to a heavy hunting pressure (See the paper by Harkonen in this volume).

2. MANAGEMENT

The legal basis for hunting in Scandinavia is that a landowner owns the wildlife on his land. The leasing of hunting rights is therefore a big market, and can be economically important to landowners. What the wildlife authorities may regulate is what species may be hunted, when hunting is allowed, how species may be hunted, and in some cases how many animals may be taken.

The Swedish system is very simple. Beaver hunting is allowed within most areas of the beavers' range. There is a hunting season from 1 October to approximately15 May. The season is longer in the northern parts of the country. In practice the hunting season is much shorter. I estimate that 90% of all beavers killed are shot within a period of a few weeks at the very end of the season. There is no bag limit on beavers in Sweden. Rifle hunting is essentially the only method used. Shotguns are prohibited, livetraps are allowed, but rarely used, Conibear traps are only allowed in extreme cases and their use is controlled by many rules, special permissions, and bureaucracy. The main reason for this is that we have a very small and sensitive otter population, and any unintended trapping of otter would be considered a catastrophe.

In Norway they are now initiating a new beaver management system based on the beaver subpopulation within a given water drainage system. The general idea is to coordinate the beaver hunting within a water drainage system and have a broader watershed management plan. The local authorities allocate quotas within the so called "Beaverregion". There is, however, the possibility for landowners to join into larger subunits and present a management plan of their own. If the plan is approved they may decide the annual harvest themselves. If this new program works it will be a very sophisticated management system. The new system may also be rather laborious and thereby expensive.

In Finland, hunters must apply for a beaver hunting license from the local wildlife management authorities, and must define how many animals they want to kill. Licenses will be granted based on the extent of the local beaver population, and the occurrence of

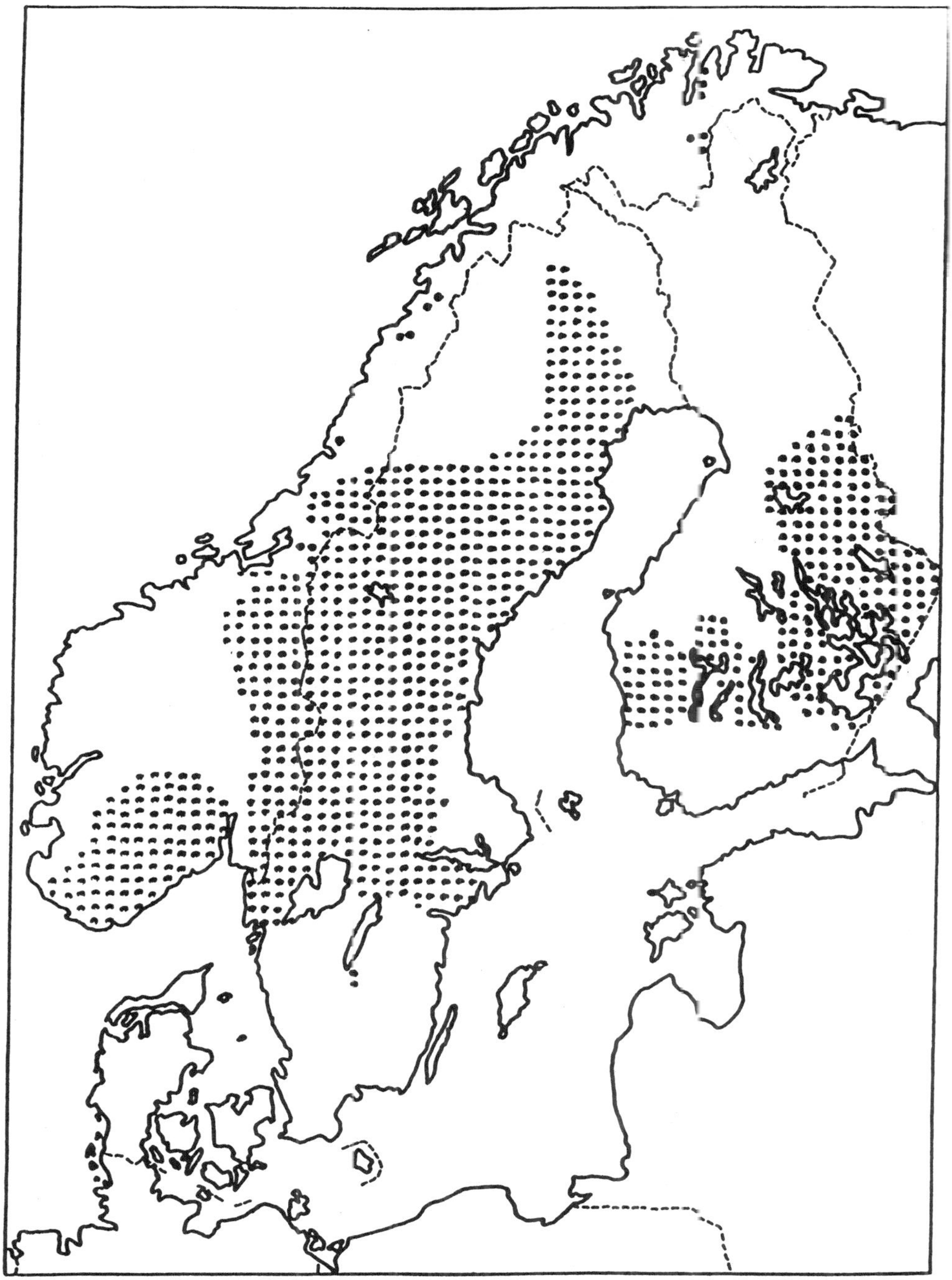

Figure 1. The distribution of beaver in Scandinavia.

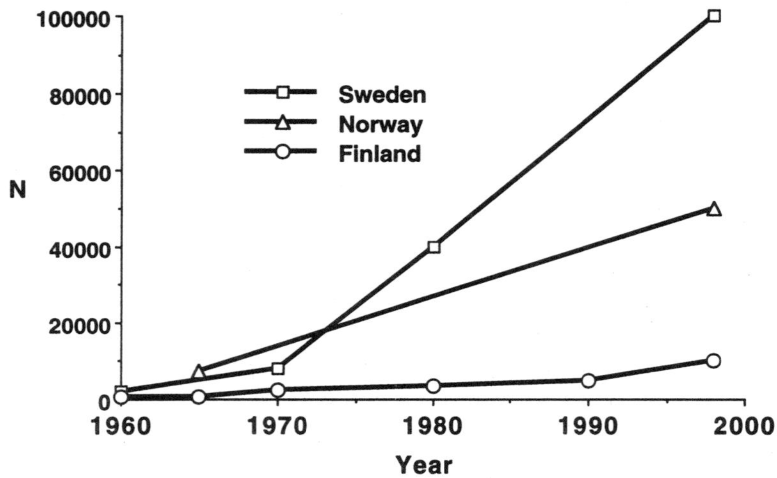

Figure 2. Beaver population development in Scandinavia.

beaver damage in the area. However, the authorities are very generous in granting licenses and it can be questioned if the annual harvest result would be different with a no bag limit system.

As I have outlined above, there are three rather different management systems presently being employed in Scandinavia, and these different systems must affect the annual harvests of beavers. However, the impact of the specific management system on the beaver harvest is not always easy to predict. For example, the lack of a bag limit in Sweden should imply very high hunting pressure in this country, while the existence of a bag limit in Finland should imply a lower hunting pressure. The examination of actual annual harvest records provides a very different situation.

2.1. Annual Harvest

The annual harvests are approximately 6000 beavers in Sweden, 5000 in Norway, and 2000 in Finland. In Sweden, the 6000 beavers taken each year represent 6% of the total beaver population, while the harvest is 10% of the population in Norway, and possibly as much as 20% of the population in Finland. The much higher percent of the total population taken in Finland is probably the explanation for the slow increase in this population (Figure 2).

2.2. Factors Affecting Hunting Pressure

When evaluating a management system it is important to take into account the possible reasons why people may want to hunt beaver. Here I will discuss four possible reasons: (1) There could be a commercial value of the meat for food and/or the pelts for fur products. (2) There could be a personal value to the individual hunter. (3) There could be a recreational value in beaver hunting. (4) Hunting could be a way of decreasing beaver damage when it becomes too disruptive.

In Scandinavia we can disregard the first point. The market for beaver pelts and meat is so small that it is negligible. There is no great economic gain for the hunter in beaver hunting. The second point is also not important. Although many hunters eat beaver meat and/or use the fur it is more a byproduct than the actual reason why they hunt. If a Scandinavian hunter wants meat in his freezer he will hunt for moose, deer, or roedeer, not beaver.

The two remaining points are the ones that matter. Most hunters that hunt beaver do it because it's fun; it is good sport. Some hunters will pay for the opportunity to hunt beavers. The intensity of the hunt will then be affected by factors as, the density of hunters, and how interested they are in beaver hunting compared with other available game species. As mentioned above, almost all beavers killed in Sweden are shot during the last two weeks of the season. The reason is that during the fall, hunters are occupied with game species considered more interesting, like moose, roedeer, hare, and various birds. During winter, when waters are ice covered, there is almost no beaver hunting because rifle hunting, not trapping, is preferred or the only legal method. Also, seasons are still open for the more interesting game species. However, during the ice-break up in the spring, beaver is almost the only game species with an open season, and this is the time when beavers are most easily shot with a rifle. So, even if the beaver season were shortened by eliminating the first five months, it would probably not make much of a difference in the total harvest.

The last point, hunting to stop or decrease beaver damage, is important. The greatest problem caused by beavers is dam building, which creates ponds, causing damage to commercial forestry. The impact damage control has on hunting pressure depends on several factors, apart from a diversity in the occurrence of damage. First of all it is question of attitude. What is the general view on wildlife, and nature versus culture? Is a forest considered to be a natural environment, or just a field where you grow timber as a crop? How much damage is then considered tolerable? My observation is that on this point we find a clear difference in beaver management between Finland and Sweden. The tolerance towards wildlife damage is much lower among Finnish landowners than among Swedish landowners.

Secondly, tolerance will differ among individual landowners. In general, large landowners will be more tolerant than small landowners. This is understandable since the same degree of beaver damage can be economically catastrophic to a small landowner, while to a large landowner it may merely be a nuisance. This means that the distribution of large and small landowners will affect the hunting pressure in an area. This is a major difference between Norway and Sweden, since the average size of properties is much smaller in Norway than in Sweden.

Thirdly, are there any other wildlife species causing problems in the area? If there are other wildlife species causing more severe damage they will get the most attention This is the case in most parts of Sweden where moose cause many times more damage than beavers. Foresters also experience problems with roedeer and small rodents. In addition to wildlife damage, more sever economic threats are falling market prices for timber increasing import of cheap timber from abroad, new rules and regulations, and taxes. Beaver damages to forest land is just one of many problems foresters encounter, and will be judged in this context.

Finally, I believe that a landowner's feeling of being free or restricted in their possibilities to affect their situation, will have an influence on their attitude and hunting behavior. If we hope to get acceptance by landowners for a species which is returning to our landscape, I believe it is important that they are given the ability to control damage themselves. It is important that they have enough control so they believe that they can control damage. My observation is that when an area is newly colonized by beavers the landowners are at first happy and actually proud about having a new, interesting, and somewhat exotic species on their land. Only when the population increases, and consequently the amount of damage increases, will they start to complain. It is during this phase of the beaver-human relationship when illegal activities, like blowing up lodges with dynamite, and destroying dams in mid-winter are observed. As soon as the area is opened for beaver hunting the complaints and illegal activities decrease or stop completely. This does not necessarily mean that the beaver population decreases, or even that the amount of damage decreases, but it is now up to the landowners to do something about the problem. Often landowners just don't think it is worth the effort, except for maybe removing a few dams. The dams are usually the direct cause of problems and it is a lot more time efficient, at least short-term, to destroy dams than to solve the problem indirectly by hunting beavers. In Sweden and Finland, in areas where beaver hunting is allowed, a landowner may destroy dams during the summer without special permission. This period is from the first of June to the last of August in Sweden, and from the 15th of June to the 15th of September in Finland. The local authorities may also give permits to destroy dams at other times of the year if it is considered necessary to prevent considerable damage. In Norway permission from the local authorities is always required.

3. CONCLUSIONS

I believe we have to be very pragmatic when it comes to beaver management. No matter how elegant and scientifically correct management plans may seem in theory, they will be of no value unless they are adapted to the current social and economic situation in a specific area, and are accepted and supported by the landowners. The Scandinavian beaver populations are still increasing and expanding their ranges. Because of this beavers exhibit an irruptive pattern illustrated by the fact that there are areas not yet colonized, areas recently colonized, areas with very high population densities, and areas with markedly declining populations (Hartman 1994). This suggests that the beaver situation in Scandinavia will certainly change in the future, and that humans must be able to change the manner in which the populations are managed. In this context we must also consider the rapid change in the forest landscape caused by beaver activities. The condition of the beaver populations, and the structure of the management systems, as well as the social and economic situations, differ markedly among the Scandinavian countries. My evaluation is that each Scandinavian beaver management system has been effective, within its specific context, and my recommendation is that effective management systems should not be altered.

REFERENCES

Aaris-Sørensen, K., 1988, Danmarks forhistoriske dyreverden. Fra istid til vikingetid. Gyldendals forlag.

Fischer-Andersen, I. B., 1997, Reintroduktion af baeveren (*Castor fiber*) til Danmark? En konsekvensanalyse. Kandidatafhandling, Univ. of Copenhagen, Museum of Zoology. 86pp.

Hartman, G., 1994, Long term development of a reintroduced beaver population in Sweden. *Conservation Biology* **8:**713–717.

Hartman, G., 1995, Population development of European beaver in Sweden, in: *Proceedings of the Third Nordic Beaver Symposium.*, (A. Ermala and S. Lahti, eds.), pp. , Finnish Game and Fisheries Research Institute, Helsinki.

Lahti, S., 1997, Development of populations, distribution problems and prospects of Finnish beaver populations, in: *Proceedings of the First European Beaver Symposium*, Bratislava, Slovakia.

Nolet, B.A., and Rosell, F., 1998, Comeback of the beaver (*Castor fiber*): An overview of old and new conservation problems, *Biological Conservation* **83:**165–173.

2

MANAGEMENT OF THE NORTH AMERICAN BEAVER (*CASTOR CANADENSIS*) ON THE SOUTH-SAVO GAME MANAGEMENT DISTRICT, FINLAND (1983–1997)

Sauli Härkönen

South-Savo Game Management District
P. O. Box 14, FIN-51901
Juva, Finland
E-mail: sauli.harkonen@riistanhoitopiirico.inet.fi

1. ABSTRACT

Two pairs of North American beavers (*Castor canadensis*) from the USA were introduced in Sääminki in southeastern South-Savo in 1937. The present population is the offspring of these two pairs. Currently, the beaver population is rapidly increasing and causing considerable damage to forest and agricultural land. Therefore there is a need to manage the beaver population at a lower, but sustainable level through hunting. However, it is impossible to develop a beaver management plan without quantitative data about beaver from this area. The South-Savo Game Management District has estimated the population density of beavers using annual questionnaires to hunters from 1983–1997. Furthermore, the occurrence and amount of damage caused by beavers was determined in 1995 and 1997. Hunting clubs manage 68% of the total area of South-Savo which encompasses about 2.17 million ha. Twenty-five percent of this area is water. The beaver population has increased from 1100 to 3000 individuals from 1983–97, and the respective number of lodges has increased from 400 to 1280. The beaver density is highest in the southeastern part of the study area, but the range is also expanding to the west. The occurrence of damage had a patchy distribution and was highest in the eastern part of the study area where the beaver density was highest. The total area of extensively flooded land caused by damming activity was 475 ha in 1995 and 321 ha in 1997 while overly wet areas increased from 581 ha to 598 ha, respectively. During the study period, 14% to 27% of the beaver population was harvested annually. Hunting is only allowed in areas where

Beaver Protection, Managment, and Utilization in Europe and North America, edited by Busher and Dzięciołowski.
Kluwer Academic / Plenum Publishers, New York, 1999.

damage to forests and agricultural lands occur. Additionally, an annual average of 32 licences allowing the destruction of problem dams have been granted from 1993–1997. In the future, management decisions for the North American beaver population in South-Savo will be based on results of annual questionnaires conducted by the South-Savo Game Management District.

2. INTRODUCTION

In Finland, beavers were hunted to extinction. The last European beaver (*Castor fiber*) was killed in northern Finland in 1868. In the mid 1880s, there were already plans for beaver reintroduction (Moilanen, 1980). However, reintroduction was delayed until the 1930s because of problems obtaining beavers. Seventeen European beavers from southern Norway were imported and reintroduced in several places in Finland in 1935 (Lahti and Helminen, 1969). Additionally, seven North American beavers (*Castor canadensis*) obtained from the USA were introduced in 1937. Two pairs of *C. canadensis* were released in Sääminki in southeastern part of South-Savo, and the present North American beaver population in Finland is composed of the descendants of these two pairs.

It was estimated that the beaver population in Finland (both species) was between 450 and 500 individuals by 1955 (Linnamies, 1956), and had increased to between 500 and 600 individuals by 1960 (Linnamies, 1961). Based on data collected through questionnaires Lahti and Helminen (1980) reported that the number of beavers had increased to between 1800 and 3000 individuals by 1965, to between 1800 and 3200 individuals by 1970, and to between 4000 and 6000 individuals by 1975. Ermala, Helminen, and Lahti (1989) estimated the number of beavers in Finland in 1980 to be between 2500 and 4400 and in 1985 to be between 2800 and 4600. At present, the European beaver population living in the western part of the country is approximately 1000 individuals, while the North American beaver population found mainly in central and eastern Finland is estimated to be 8000 individuals (Ermala, 1996). These population sizes are rough estimates based on the responses to questionnaires.

In Finland, damage to forests has increased as the beaver population has increased. Lahti and Helminen (1969) reported 800 ha of conifers killed by flooding and an additional 1900 ha of forest affected by higher water levels in 1965. By 1975 these damage estimates had increased to 2000 ha of trees killed and 2400 ha damaged (Lahti and Helminen, 1980). Damage similar to this has been reported in Lithuania and Estonia which also have increasing beaver populations (Laanetu, 1995; Mickus, 1995). Beavers also fall and debark valuable trees which can cause considerable economic loss for landowners. In Finland, damage caused by beavers is not compensated by the government unlike moose (*Alces alces*) damage for which the government does provide compensation.

North American beaver hunting was initiated in the 1960s in response to the increasing damage reports. European beavers were not hunted until the 1980's (Ermala *et al.*, 1989). Annual harvests of both species has steadily increased (Ermala, 1996). Hunting is carried out on a license basis, but the open season has varied. At present, the open season is from August 20 to April 30, and the number of licenses is restricted and controlled by local Game Management Districts. Hunting licenses are directed to areas where damage to forests and agricultural lands occur. Also licenses allowing the destruction of problem dams have been granted.

Successful management of beaver populations is impossible without quantitative data on population demographics and damage. The present study was designed to: 1) estimate the population size, distribution and density of North American beavers, 2) determine the occurrence and degree of beaver damage, and 3) gather and analyze the hunting statistics in South-Savo Game Management District from 1983 to 1997.

3. METHODS

The South-Savo Game Management District (SGMD) is in southeastern Finland (Figure 1). The SGMD consists of 25 Game Management Units. The total area of SGMD is about 2.17 million ha, of which 25% is water. Lakes and rivers in the study area provide very good beaver habitat with an estimated 38,000 km of land/water edge in SGMD. Forests cover 80% of the land area and Scots pine (*Pinus sylvestris*) is the dominant tree species (40% of all trees). Deciduous trees account for approximately 20% of the total tree species on SGMD.

From 1983 to 1997 SGMD has estimated the number of lodges and the population density of North American beavers (minimum and maximum estimates) using annual questionnaires to hunters (Table 1). Questionnaires were sent to the local hunting clubs in September of each year, since these hunting clubs manage 68% of the total area of SGMD. Additionally, the occurrence and amount of damage to forests and agricultural lands caused by beavers was examined in 1995 and 1997. Damage was divided into extensively flooded areas (i.e. trees were dead or are dying) or overly wet areas (i.e. trees were suffering and the growth is diminished).

Hunting statistics gathered by SGMD were used to determine annual harvest rates, and the number of licenses allowing destruction of problem dams was also examined.

4. RESULTS

The beaver population has increased from 1100 to 3000 individuals in SGMD from 1983 to 1997, and the number of lodges has increased from 400 to 1280 respectively (Figure 2). The beaver density (individuals/1000 ha) was highest in the southeastern part of the study area during the study period (Figure 3). The beaver population is, however, expanding its range to the west.

The occurrence of damage had a patchy distribution, but was highest in the eastern part of the study area which had the highest beaver density (Fig 3). The total area of extensively flooded land caused by damming activity was 475 ha in 1995, and an additional 321 ha was flooded in 1997. Overly wet areas caused by beaver activity were 581 ha in 1995 and 598 ha during 1997. The decrease in the annual area of extensively flooded land may due to the increasing hunting harvest in damage areas.

Beaver hunting licenses were directed to areas where damage to forests and agricultural lands occurred. The allowed hunting harvest has increased from 491 beavers in 1983 to 999 beavers in 1997 (Figure 4). The number of beavers taken (bag) has been between 36% and 60% of the allowed limit. During the study period 14% to 27% of the beaver population was harvested annually. In 1985, hunting was allowed in 15 of 25 Game Management Units, and by 1997, hunting was allowed in 22 Units. Furthermore, an annual average of 32 licenses allowing destruction of problem dams were granted from 1993 to 1997.

5. DISCUSSION

Recently, a rapid increase in the beaver population has occurred in SGMD. The present SGMD beaver population is more than 30% of the total North American beaver popu-

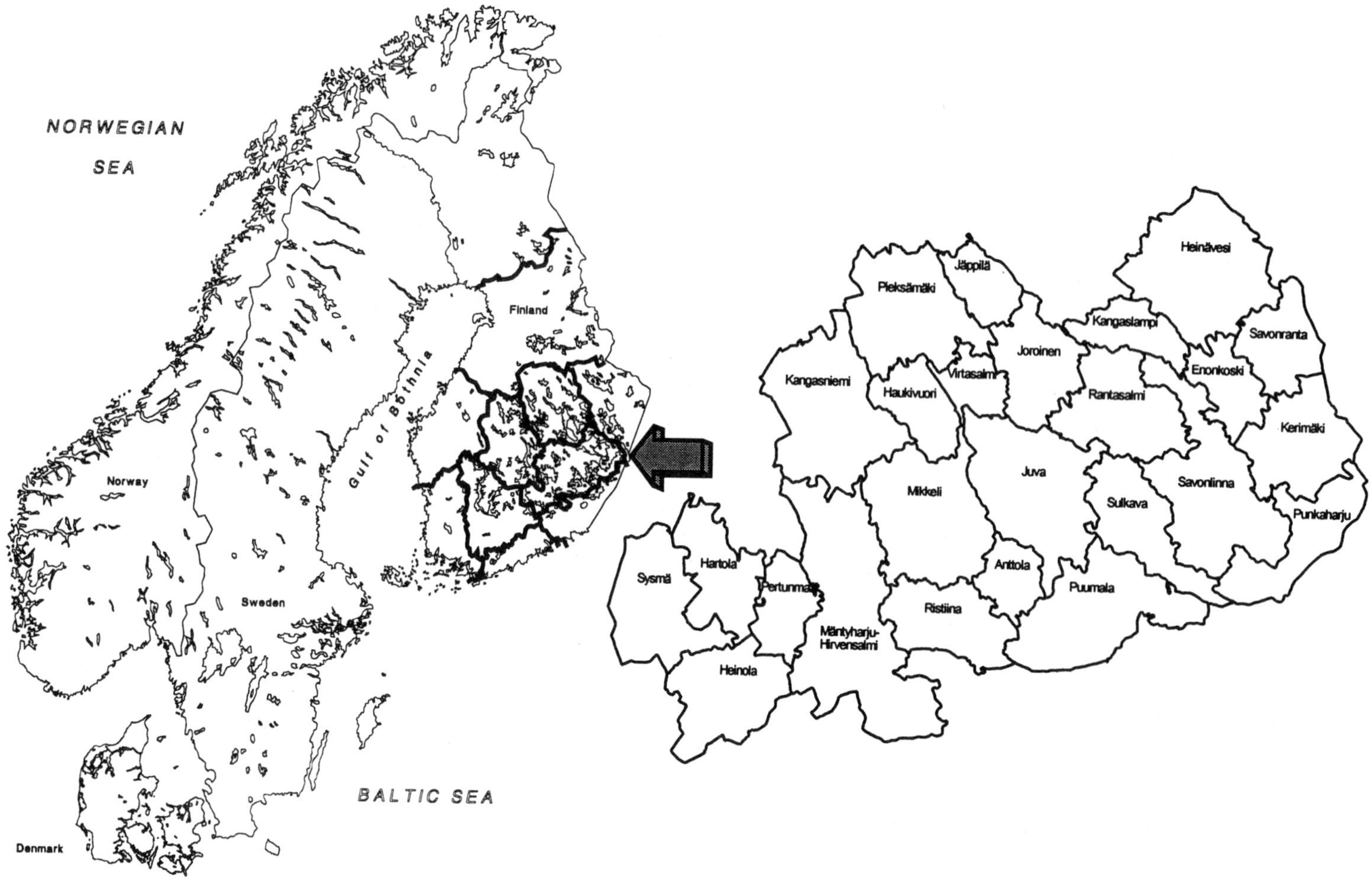

Figure 1. The study area in the area of South-Savo Game Management District in southeastern Finland. South-Savo Game Management District has 25 game management units.

Table 1. Questionnaire sent to all moose hunting clubs in the area of South-Savo Game Management District

South-Savo Game Management District			
Beaver questionnaire	19________		
Game management unit	________		
Hunting club	________		
Area of hunting club	________ ha		
Number of beavers			
No occurrence of beavers	☐		
Estimate	________ minimum		________ maximum
Change in beaver population (in relation to previous year)			
	increased ☐	decreased ☐	no change ☐
Number of lodges	________		
Damage caused by beavers			
Occurrence of damage	☐ yes	☐ no	Total area ________ ha
Damage type	extensively flooded area	________	ha
	overly wet area	________	ha
Date and signature:	________		

lation in Finland (c.f. Ermala, 1996). However, since the population size in SGMD is only an estimate, the actual population size may be even larger. Thus, it is important to gather as accurate data as possible for management purposes.

In the SGMD the number of beavers is still increasing and there are still many areas available for beavers to colonize. Because of the available habitat I estimate it will take a number of years for the beaver population to reach the carrying capacity of the SGMD. At present, hunting is the main mortality factor, which can be quantified. Severe winters can also affect the beaver population, but there is at present no quantifiable data on climate induced mortality. In addition, there are some observations in the study area that wolf (*Canis lupus*) and lynx (*Lynx lynx*) have killed beavers during spring. At present this predation is negligible.

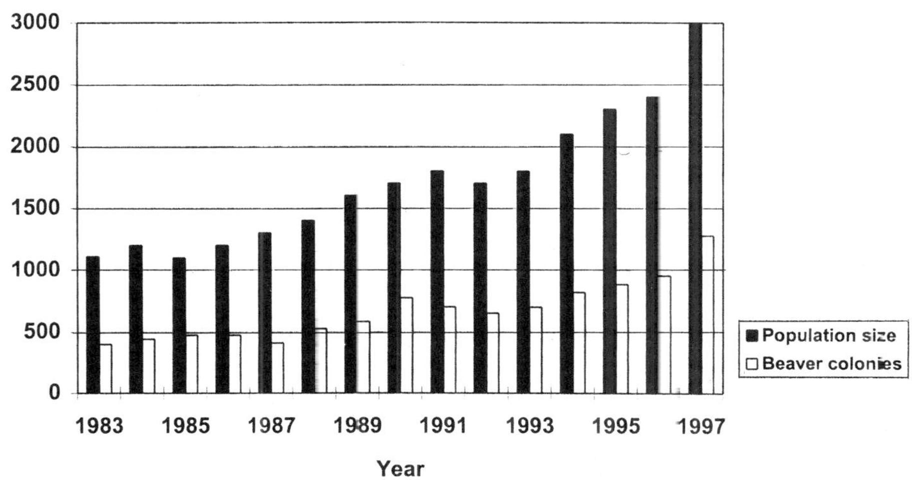

Figure 2. The estimated population size and the estimated number of beaver colonies in South-Savo Game Management District during the period 1983–1997.

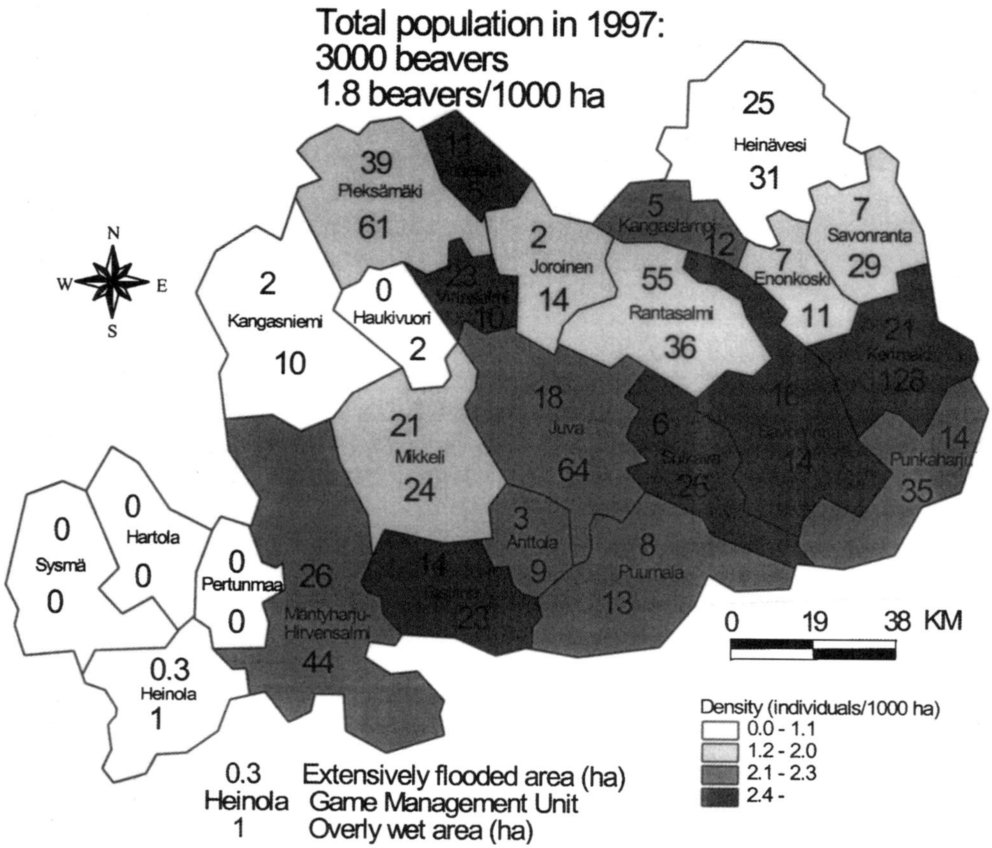

Figure 3. The beaver density (individuals/1000 ha), area extensively flooded by damming (ha), and area overly wet by damming (ha) in South-Savo Game Management District in 1997.

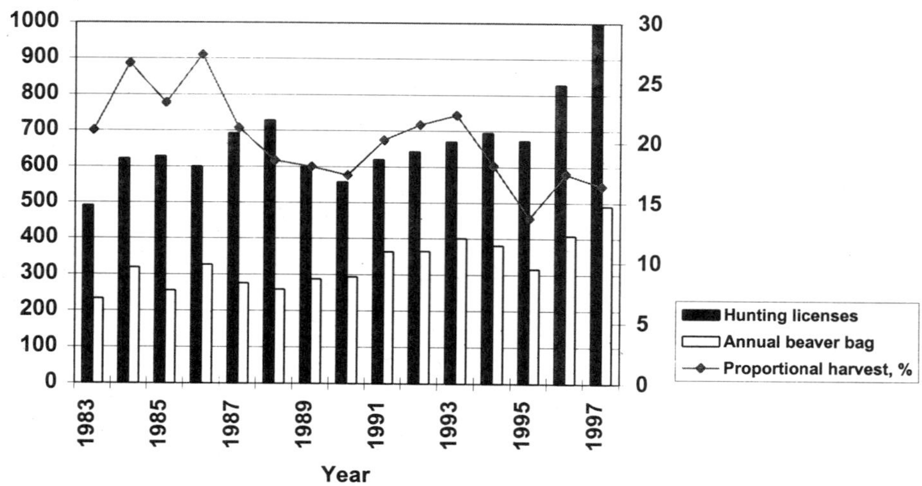

Figure 4. The allowed hunting harvest (number of hunting licenses), annual beaver bag (number of beavers taken), and proportional harvest (%) (beaver bag/population size*100) in South-Savo Game Management District during the period 1983 to 1997.

Evidence that the value of beaver as a game animal has decreased is that the number of beavers taken during the season (bag) has only been approximately 50% of the allowed limit. In the past, beaver fur was very valuable and the meat was also used. Currently, beaver fur has no commercial value and the majority of hunters do not like to eat beaver meat. The relatively low harvest rate may also illustrate that it is quite difficult to hunt beavers. Thus, even with increasing hunting pressure the beaver population has continued to increase.

Damage caused by beavers to forests and agricultural lands was relatively low in relation to the total land area. Forest and land damage caused by beavers can be considerable, especially for private landowners, because damage occurs in a patchy manner. In SGMD, private forest-owners (30,000) manage 80% of the forest area. Thus, it is important to take into account the occurrence of damage when directing hunting licenses to the right areas. In the areas where hunting is not needed the licenses allowing destruction of problem dams have been granted. At present, I have no figures about the economic loss caused by damming. Felling and debarking of trees are also considered types of damage. The total value of the damage could be calculated by using the forest type, age category, volume and tree height. I have already inspected 50 damage areas and will continue this work which I will publish later.

Beavers can effectively affect landscape. Deciduous trees such as aspen (*Populus tremula*) and willows (*Salix* spp.) are favored in the diet. Selective harvesting of deciduous trees by beavers can increase the succession process towards the coniferous phase (Naiman, Johnston, and Kelley, 1988). Beavers can also bring the succession to the earlier phase when felling of deciduous trees is directed to the canopy trees (Pastor and Naiman, 1992). In such cases, openings are created and young seedlings can be naturally regenerated. In addition, beavers have been shown to have long-term influences on the forest landscape which has been studied on the Kabetogama Peninsula, Minnesota, USA between 1927–1988 (Naiman, Pinay, Johnston, and Pastor, 1994). In this area, initially there were few beaver ponds, but by the end of the study period beaver ponds covered 13% of total peninsula area. In my study area, the beaver density is still so low that enormous landscape effects are not seen. I do observe gradual changes in landscape, such as an increase in the patchiness of habitat types, but more long-term data are required.

In the study area I have observed that public opinion is very tolerant regarding the presence and activities of beavers, especially during the early stages of colonization. Beavers and their activities are interesting to observe, and aesthetically pleasing to most people. However, the intensive damming activity dramatically changes this situation. Beavers rapidly are viewed as a damaging agent, and when this occurs the role of the local Game Management Districts is very important. Local game managers must have the capability to respond rapidly by allowing controlled hunting.

Information on beaver population size, density, and range expansion, coupled with accurate estimates of beaver damage, allows game managers to reconcile the ecological benefits provided by beavers with the economic damage caused by their activities. This study indicates that restricted hunting of beavers has been effective in controlling the impact of their activities on forest lands, while not having a negative effect on the population. On the SGMD I am observing a relatively small amount of forest damage, and, at the same time, an increase in the beaver population size. These factors suggest that the SGMD beaver management plan is a success, and is providing benefits for both beavers and humans.

ACKNOWLEDGMENTS

I would like to thank Dr. P. E. Busher for constructive and very helpful comments on the manuscript. I also wish to thank Dr. G. Hartman who encouraged me to write something about the North American beaver in Finland. I am also grateful to the hunting clubs for their responses to the questionnaires. I also thank the South-Savo Game Management District for providing the facilities to complete this study.

REFERENCES

Ermala, A., 1996, Euroopanmajava (*Castor fiber*), kanadanmajava (*Castor canadensis*), in: *Riistan jäljille*, (H. Linden, M. Hario, and M. Wikman, eds.), pp. 30–33. Riista- ja kalatalouden tutkimuslaitos, Edita, Helsinki. (In Finnish with English summary).

Ermala, A., Helminen, M., and Lahti, S., 1989, Majaviemme levinneisyyden ja runsauden vaihteluista sekä tulevaisuuden näkymistä (Summary: Some aspects of the occurrence, abundance and future of the Finnish beaver population), *Suomen Riista* **35:**108–118.

Laanetu, N., 1995, The status of European beaver (*Castor fiber* L.) population in Estonia and its influence on habitats, III Nordic Beaver Symp., 1992 Ilomantsi, Finland, pp. 34–40.

Lahti, S., and Helminen, M., 1969, Suomen majavien istutushistoriasta ja kannan levinneisyys 1960-luvulla (Summary: History of reintroductions and present population status of the beaver in Finland), *Suomen Riista* **21:**67–75.

Lahti, S., and Helminen, M., 1980, Suomen majavien levinneisyyden muutokset vuosina 1965–1975 (Summary: The status of European and Canadian beavers in Finland in 1965–75), *Suomen Riista* **27:**70–77.

Linnamies, O., 1956, Majavien esiintymisestä ja niiden aiheuttamista vahingoista maassamme, *Suomen Riista* **10:**63–86. (In Finnish).

Linnamies, O., 1961, Majavakantamme viimeaikaisista muutoksista (Summary: Recent changes in Finnish beaver (Castor) population), *Suomen Riista* **14:**159.

Mickus, A., 1995, The European beaver (*Castor fiber* L.) in Lithuania, III Nordic Beaver Symp., 1992 Ilomantsi, Finland, pp. 44–45.

Moilanen, P., 1980, Täydentäviä tietoja Suomen majavien istutushistoriaan (Summary: Additions to the history of beaver reintroduction in Finland), *Suomen Riista* **27:**78–80.

Naiman, R. J., Johnston, C. A., and Kelley, J. C., 1988, Alteration of North American streams by beaver, *BioScience* **38:**753–762.

Naiman, R. J., Pinay, G., Johnston, C. A., and Pastor, J., 1994, Beaver influences on the long-term biogeochemical characteristics of boreal forest drainage networks, *Ecology* **75:**905–921.

Pastor, J., and Naiman, R. J., 1992, Selective foraging and ecosystem processes in boreal forests, *American Naturalist* **139:**690–705.

3

REINTRODUCTION OF THE BEAVER, *CASTOR FIBER*, IN THE NETHERLANDS

Vilmar A. A. Dijkstra

Dutch-Belgian Mammal Society (VZZ)
Oude Kraan 8
6811 LJ Arnhem, The Netherlands

ABSTRACT

The European beaver, *Castor fiber*, was hunted to extinction in The Netherlands in the 1800s. However, suitable habitat is still present, and beavers were reintroduced between 1988 and 1991 in the National Park de Biesbosch. The introduced animals suffered high mortality and the population grew very slowly. The total population size in 1997 was 65 animals and it is increasing at a very slow rate. A potential deficiency in sodium and phosphorus may affect the ability of pairs to reproduce, and could influence the territory size. The Biesbosch region is also contaminated by heavy metals, such as cadmium, and this may also influence beaver reproduction in the future.

1. INTRODUCTION

In the 19th century the European beaver, *Castor fiber*, became extinct in The Netherlands due to hunting. There were several reasons for reintroducing the beaver into The Netherlands: (1) suitable habitat was still present, (2) the positive influence of the beaver on vegetation structure and diversity along waterbanks, (3) the cause of extinction was no longer relevant, and (4) spontaneous settlement was not likely.

Between 1988 and 1991 a total of 42 beavers were released into the National Park de Biesbosch in The Netherlands. Reintroduction was carried out by the Institute for Forestry and Nature Research (IBN). In 1995 the Dutch-Belgian Mammal Society (VZZ) joined the research and began monitoring the population.

Beaver Protection, Managment, and Utilization in Europe and North America, edited by Busher and Dzięciołowski.
Kluwer Academic / Plenum Publishers, New York, 1999.

2. REINTRODUCTION AND POPULATION GROWTH

In the first year (1988) three pairs of beavers were released and they all survived. The second and third year more individuals were released, however, mortality was high; 26% and 21% respectively. The mortality occurred mainly among the newly released animals. The first four years reproduction was low with only 28% of the pairs reproducing (SD=5.8, 0.5 juvenile per pair present, SD=0.17, N=4). The population increased only due to the continued release of new animals. After the first four years, mortality decreased (11%, SD=2.4, N=6) and reproduction increased (55% of the pairs reproducing, SD=14.7, 1.0 juvenile per pair present, SD=0.28, N=5). In 1997 the population consisted of approximately 65 animals.

3. DISCUSSION

Although the Biesbosch population is increasing, the growth rate is relatively slow. The slow growth rate is most likely due to the low percentage of pairs that is currently reproducing. Nolet, Van der Veer, Evers, and Ottenheim (1995) calculated that the beavers in the Biesbosch could suffer a shortage of sodium and phosphorus during gestation. Sodium and phosphorus can be found in relatively high amounts in aquatic plants like *Nuphar lutea*. These plants are rare in the area and this may be the reason beaver territories in the Biesbosch are large. In 1993 the mean territory size was 12.5 kilometers of water bank (Nolet and Rosell 1994). There are no indications that the territories are getting smaller as the population increases. Possibly beavers in the Biesbosch need these large territories to satisfy their mineral requirements during gestation. This could also be the reason why the percentage of reproducing pairs is relatively low.

The Biesbosch is also highly polluted with heavy metals. Previous research (Nolet, Dijkstra, and Heidecke, 1994) revealed that the bark of willow (*Salix* Spp.), the main food of beavers in the Biesbosch, contained relatively high concentrations of cadmium (geometric mean=11 mg/kg dry weight). The kidneys of beavers born in the area and of beavers that were released and have lived there for at least 5.5 years, contained high amounts of cadmium (geometric mean=92 mg/kg dry weight, range: 18–486, N=7). A strong relationship has been found between residence time in the Biesbosch and cadmium concentration in the kidneys (P<0.001, R^2_{adj}=0.915, N=7; unpublished data, V.A.A. Dijkstra & S. Broekhuizen). Although there is no clear indication that reproduction is influenced negatively by the high cadmium concentrations it also can not be entirely discounted. It is possible that heavy metal contamination will be a significant factor influencing the Biesbosch beaver population in the future.

REFERENCES

Nolet, B. A., and Rosell F, 1994, Territoriality and time budgets in beavers during sequential settlement, *Can. J. of Zool.* **72**:1227–1237.

Nolet, B.A., Dijkstra, V. A. A., and Heidecke, D, 1994, Cadmium in beavers translocated from the Elbe river to the Rhine/Meuse estuary, and the possible effect on population growth rate, *Archives of Environmental Contamination and Toxicology* **27**:154–161.

Nolet, B. A., Van der Veer, P. J., Evers, E. G. J., and Ottenheim,M. M., 1995, A linear programming model of diet choice of free-living beavers, Netherlands *J. of Zool.* **45**:317–335.

4

THE BEAVER IN RUSSIA AND ADJOINING COUNTRIES

Recent Trends in Resource Changes and Management Problems

Alexander P. Saveljev and Vladimir G. Safonov

The All-Russian Research Institute of Game Management and Fur Farming
Kirov, Russia

ABSTRACT

In many areas of the former Soviet Union people adapted to post-Soviet era reforms by overexploiting natural resources. The European beaver, *Castor fiber*, represents a wildlife resource that was depleted in the years following perestroika. Because of this we wanted to document the current status of the beaver population in the Russian Federation and adjoining countries, examine conservation efforts, and where appropriate, investigate the effectivness of harvesting beavers as a management tool. Data are derived from official statistics, publications, questionnaires, and field studies in Russia, Belarus, the Ukraine, and Kazakhstan. At present beavers are found in 60 of 87 regions of Russia, 6 regions of Belarus, 20 regions of the Ukraine, and 3 regions of Kazakhstan. In 1997 the estimated number of beavers in these countries was 232,000, 24,000, 6,000, and 1,000, respectively. However, beavers are not distributed evenly throughout the countries. For example, more than 75% of all Russian beavers are concentrated in 14 taiga regions from Karelia to the Ural Mountains. The decrease in beaver numbers observed immediately after perestroika has stopped in the last few years, and since 1996 there has been an average annual increase of between 6% and 20%. Beaver harvesting occurs in 40 regions of Russia, and in 1997/98 a fee for using the fur resource was implemented. The range of beavers is expanding in south Siberia and Kazakhstan, and North American beavers, *Castor canadensis*, have either been introduced into, or immigrated to northern regions. In the Far East Region only North American beavers are found in two of the three subdivisions. We discuss a multi-purpose management model that includes management of genetic (subspecific) resources, preservation, and regulated harvesting. Management is necessary to main-

Beaver Protection, Managment, and Utilization in Europe and North America, edited by Busher and Dzięciołowski.
Kluwer Academic / Plenum Publishers, New York, 1999.

tain the beaver population in Russia and insure the survival of this biologically important species.

1. INTRODUCTION

Populations of European beaver (*Castor fiber* L.) have recovered from near extirpation in the European part of the former USSR and in West Siberia. Because of the increase in both the number of beavers and the distribution of beavers it was possible to restore a limited exploitation of this valuable furbearer. Limited hunting was implemented in the mid 1960s after a 300–400 year period. From the 1970s through the 1980s, within the forest zone of eastern Europe from Belorussia to the Ural mountains, the beaver was one of the main species harvested for fur. Under the conditions of delayed economic reforms (1987–1998) the disintegration of the former system of game management coupled with decreasing state control of the environment led to the increased exploitation of natural resources. This increasing exploitation affected the Russian beaver population.

The beaver represents a resource that was almost completely eliminated from Russia in the past, was restored after many years, and now may be threatened again. Because of this we believe it is important to analyze the population dynamics and the present status of beavers in Russia and adjoining countries. We will also attempt to estimate the management prospects of all the subspecies of European beavers living in the territory of Russia, Belarus, the Ukraine, and Kazakhstan. We also consider the management situation for North American beavers in this region.

2. METHODS

Data on beaver numbers were taken from official statistics. Since the official statistics were often incomplete and inaccurate, we also collected information from scientific publications, questionnaires, and our personal observations and census. When estimating populations that are being managed and preserved we took into account genetic data that we obtained during the recent years.

3. RESULTS

In the early 1990s a decrease in beaver numbers was observed in certain areas, and was most significant in the central and northwestern regions. Since the mid 1990s the situation has improved and most populations have stabilized or increased (Table 1). The beaver numbers in each economic region (broken down by administrative and territorial subdivisions) of the Russian Federation for 1995 are given in Table 2. In this table the data are from Borisov (1996), with our revisions and corrections for certain regions. Beavers inhabit 60 of 87 autonomous republics, territories and regions of the Russian Federation, all 6 regions of Belarus, 20 of 25 regions of the Ukraine and 3 regions of Kazakhstan. More than 75% of all Russian beaver resources are concentrated in 14 regions and republics of the taiga zone of the European part of the Russian Federation. The highest beaver numbers are in Kirov in the Volga-Vyatka region. In Kirov there are 21,300 individuals and an average density of 1.8 beavers per 10 km^2.

Table 1. Beaver numbers (thousands of individuals) in the states of the former USSR

	Year		
State	1995	1996	1997
Russia	209.6	219.0	232.5
Belarus	19.3	18.8	23.7*
Ukraine	no data	no data	6.0*
Kazakhstan	no data	no data	1.0*

*These are best estimates based on available data.

Regular beaver harvesting takes place in 40 regions of Russia. Since the 1997/98 season a fee for the use of fur resources was implemented in Russia. In a number of regions (Astrakhan, Buryatia, Irkutsk, Murmansk) *Castor fiber* L. is not harvested since the populations are too small (Laptenok and Savchenko,1997). In the Republic of Tyva and in the Khanty and Mansi national areas of the Tyumen region, small populations of the aboriginal subspecies of *C. fiber tuvinicus* Lavrov and *C. fiber pohlei* Serebr. are protected and entered in the IUCN Red Data Book.

On the Kamchatka Peninsula and in Primorie, North American beavers, *Castor canadensis* K., are present and are still protected. In Karelia and the Leningrad region, where beavers have settled after migrating from Finland, harvesting is permitted. In Belarus beaver harvesting has been prohibited for the past 3 years, and this has yielded positive results (Table 1). In the Ukraine there is no regular beaver harvest, but 50 to 60 individuals are taken yearly by special permit. In Kazakhstan beaver numbers continue to increase slowly. Beavers migrated to the Ural River tributaries in this region from Russia. This population has recovered to such an extent that in 1996 the European beaver was removed from the list of endangered species in the National Red Data Book of Kazakhstan (Bekenov, 1997).

From the 1950s through the 1970s beavers were actively reintroduced in many regions. During this period over 15,000 European and more than 800 North American beavers were settled. By the 1990s this reintroduction program was almost completely discontinued. In the 1990s beaver reintroduction was carried out only on the initiative of the All-Russian Research Institute of Game Management and Fur Farming, and by its research workers in the Republic of Tyva (a native subspecies *C.f. tuvinicus*). Reintroductions were also conducted in the Tomsk region, and in the Republic of Komi where hybrids of Voronezh and Belorussia beavers from Kirov were released. In total 76 animals were released. The potential for additional beaver reintroductions in unpopulated areas of northern Russia and Siberia exist, however, given the current economic conditions will most likely not occur.

4. DISCUSSION

Beavers represent a model species for analyzing management practices of renewable biological resources. There are numerous ways that beaver biology can be approached in a management context. For example, beavers represent an historic natural resource that was removed from the landscape by humans, and then became the subject of protection and conservation. This aspect can be investigated from both scientific and sociological directions. The reintroduction of this species into its historic area has also allowed the ecosystem to recover a functional link in its structure. Beavers can also be studied as an

Table 2. The status of beavers in the major regions and administrative subdivisions of the Russian Federation in 1995

Administrative subdivisions of the Russian Federation	Number of beavers (in thousands)	Exploitation	Notes
Northern Region	39.5		
Arkhangelsk	9.0	+	
Vologda	16.0	+	
Karelia	7.0	+	two species**
Komi	7.5	+	
Murmansk	rare	–	in the Red Data Book***
Northwestern Region	31.0		
Leningrad	10.0	+	two species
Novgorod	11.0	+	
Pskov	10.0	+	
Baltic Region	2.5		
Kaliningrad	2.5	+	
Central Region	37.0		
Bryansk	0.5	+	
Vladimir	1.8	+	
Ivanovo	4.5	+	
Kaluga	1.0	-	
Kostroma	10.0	+	
Moscow	1.0	+	
Orel	0.9	-	
Ryasan	1.1	+	
Smolensk	2.0	-	in the Red Data Book
Tver	10.0	+	
Tula	0.7	+	
Yaroslavl	3.5	+	
Volga-Vyatka Region	27.8		
Kirov	21.3	+	
Mari-El	2.0	+	
Mordovia.	0.5	+	
Nizhm Novgorod	3.0	+	
Chuvashia	1.0	+	
North Caucasus Region	0.4		
Rostov	0.4	-	
Central-Chernozern Region	5.2		
Belgorod	0.2	-	
Voronezh	2.5	+	
Kursk	0.8	+	
Lipetsk	0.5	-	
Tambov	1.2	-	
Povolzhie Region	10.1		
Astrakhan	rare	-	
Volgograd	3.0	+	
Pensa	1.0	-	
Samara	0.8	-	
Saratov	3.0	+	
Tatarstan	1.0	+	
Ullanovsk	1.3	+	
Ural Region	35.5		
Bashkortostan	1.5	+	
Kurgan	1.0	-	
Orenburg	7.0	+	
Perm	7.0	+	
Sverdlovsk	7.0	+	

Table 2. (*Continued*)

Administrative subdivisions of the Russian Federation	Number of beavers (in thousands)	Exploitation	Notes
Udmurtia	10.0	+	
Chelyabinsk	2.0	+	
West-Siberian Region	12.9		
Altai	4.0	+	
Kemerovo	1.5	+	
Novosibirsk	1.5	+	
Omsk	1.4	+	
Tomsk	1.0	+	
Tyumen	3.5	+	in the Red Data Book, one subspecies
East-Siberian Region	94		
Buryatia	rare	–	
Irkutsk	0.8	–	in the Red Data Book
Krasnoyarsk	8.5	+	
Tyva	0.1	–	in the Red Data Book
Khakasia	0.2	–	
Far East Region	0.6		
Kamchatka	0.2	–	only North American beaver
Primorie	rare	–	only North American beaver
Khabarovsk	0.4	+	two species
Total	209.6		

+ Regular harvesting takes place
– Harvesting is prohibited
**Both *Castor fiber* L. and *C. canadensis* K.
***Entered in a regional or federal Red Data Book

introduced species in the ecosystem since they were an absent component for many years. The function of beavers in the ecosystem can thus be evaluated. Since there are regions where the two beaver species are sympatric, the interactions and interrelationships of these two species can be investigated. Beavers also represent a potential resource for fur products and this aspect can also be considered. Because of the numerous ways beavers can be viewed it is important to differentiate the following aspects of management.

4.1. Evolutionary and Genetic Management

Any system of population management of a species of game animal that includes harvesting and active introductions is defined to a great extent by the evolutionary fate of that species. This includes intraspecies differentiation, the adaptive potential of subspecies, and the selection of the best adapted subspecies for introduction. When developing and implementing resource management programs little attention has been given to how this aspect will influence the effectiveness of the program.

The North American beaver is evolutionarily younger, and apparently more competitive than the species from Eurasia. Observations in areas where the beaver species are sympatric (in Finland and in northwestern Russia), indicate that it is necessary to separate the species or the North American beaver will become dominant (Ermala, Helminen, and Lahti, 1989; Nummy and Pienmunne, 1998). It is also necessary to avoid the potential for interspecies overlap when carrying out deliberate introductions of *C. fiber*.

Only three of eight *C. fiber* L. subspecies are numerous (Norwegian, Belorussian, West-European) and the population numbers of the other five subspecies remain low. A higher genetic potential and variability level of the West-European subspecies from the Voronezh preserve and from the Belorussian preserve in the Berezina river basin, was discovered. A hybrid population from the Kirov region (the Cheptsa river) considerably exceeded the allelic diversity observed in both of the above mentioned ancestor populations (Milishnikov, Likhnova, and Nikonova, 1994; Milishnikov, Saveljev, and Likhnova, 1997). This is evidence that a management strategy must be chosen that takes into account the features of specific populations (interspecies differentiation, adaptive and breeding advantages).

In practice both in Russia and in western Europe both beaver species (*C. fiber* and *C. canadensis*) were introduced, and subspecies within a species were inevitably crossed. Present day populations are thus predominantly hybrid. The behavior of humans towards different beaver subspecies has been variable. For example, in eastern Germany (the Schwerinsk Lakes) during 1986–1989, an artificially created population of melanistic beavers of Voronezh origin was removed in order to preserve the purity of the Elba subspecies (Heidecke and Dornbusch, 1990). However, in western Germany, Switzerland, and Austria the attitude towards non-native beaver subspecies has been more open. In the past even populations of North American beavers have been tolerated in these regions.

4.2. Demographic Management

Beaver population dynamics depend on the local population density. Therefore, environmental factors may influence the fluctuation of beaver population sizes to a lesser degree. We believe that to a large extent beaver population size is self-regulated (population homeostasis). Limited territories suitable for colonization, and deficiencies in quantity and quality of food plants help stimulate the mechanism of self-regulation. In high density situations the typical family structure may be altered. Average group size increases, and young adult animals (two-year-olds) may delay dispersal (Busher 1987; Busher, Warner, and Jenkins, 1983; Lyons and Busher this volume). As density increases so does agonistic behavior between beavers, which can be documented by an increase in the number of observed bite marks on the tails. Under these conditions the percent of sexually mature, non-breeding females increases and the age of sexual maturity may be delayed. An increase in the embryonic and post-embryonic death rate may also be observed. Together with these processes the best indicator of the beginning of population self-regulation is a decrease in the annual growth rate. The slowing in the annual rate of increase during a period of population growth may serve as a signal for managers to begin considering active exploitation. The decrease in the annual rate of population growth was noticed for the first time on the Kerzhenets river in the Nizhni Novgorod region (Gudkova, 1954). This aspect of beaver demographics was later documented by many other researchers in other territories, and the significance of this phenomenon was used as the basis for initiating regulated beaver harvesting (Safonov, 1962; Safonov and Semyenov, 1962; Zharkov, 1968).

Novak, Baker, Obbard, and Malloch (1987) summarize the management experience with North American beavers, and note the importance and advisability of timely exploitation in making rational use of the beaver resource. They stress the use of the correlation between current population size and harvest size in beaver management programs in North America. In Russia this model of resource management was not followed, or it occurred too late, and the result was the useless loss of animals.

4.3. Regional Management Concerns

Both scientists and non-scientists are aware that beavers alter their habitats, change the hydrologic regime of aquatic systems, and influence the quality and quantity of the riparian vegetation. Management of the beaver resource must take into account regional differences in the beaver population, and in the landscape. In regions where beaver numbers and density are high, priority should be given to harvesting beavers in an attempt to control the population. In other regions where beaver numbers are low they should be protected, and any control effort should take into account the ability of beavers to carry out ecosystem services in wetlands. In some regions it may be necessary to actively control beaver numbers to prevent damage to aquatic systems. Our experience is that one management model can not be equally effective in all regions. While the model suggested by Balodis (1989) is designed to attain general ecological aims it is not multi-purpose. For this reason it may only be applicable to the beaver situation in Latvia.

4.4. Economic Management

After successful reintroduction programs in North America, Europe, and Russia the beaver populations have continued to grow in these regions. As the beaver populations increased so did human exploitation of this resource. In Canada and the United States up to 1,000,000 (one million) beavers, may be taken annually, and in Russia the number is approximately 30,000 (thirty thousand). Regulated harvesting, started at the proper time to eliminate homeostatic population control, will allow a population to realize its full reproductive potential. In such a situation both economic and ecological aims may be attained. Regulated harvesting is considered a promising management tool, and apparently has been the traditional way this resource has been exploited, especially in the northern territories by indigenous peoples.

5. CONCLUSIONS

The methods of intensive beaver management are known, but they are not yet fully realized for either economic or ecological purposes. Natural resources in the countries of central and western Europe necessary for increasing beaver numbers are limited. For the purpose of maintaining both biological diversity and genetic diversity, beaver preservation and breeding programs are currently being carried out. These programs will continue in the future to insure the survival of this important biological component of our natural landscape.

ACKNOWLEDGMENTS

The authors are grateful to Mr. V. Kwakin (Minsk), Mr. O. Dudkin (Kiev) for the information they provided, and also to Mrs. T. Vasenina (Kirov) for translation of the manuscript from Russian into English. We also thank Mrs. L. Dolgoarshinnykh (Kirov) for preparing the manuscript.

REFERENCES

Balodis, M. M., 1989, Biological principles of beaver management (Latvia as an example), Author's abstract of the thesis submitted for a degree of Doctor of Biology, Moscow, p. 1–40.

Bekenov, A., 1997, Rare species of Kazakhstan mammals and the problems of their protection, in: Rare species of Russian mammals of the adjacent territories, Moscow, p. 10.

Borisov, B. P., 1996, The beaver. The resources of the main species of game animals, and hunting grounds of Russia (1991–1995), Hunting Department of Russian Federation, Moscow, p. 137–140.

Busher, P.E., 1987, Population parameters and family composition of beaver in California, *J. Mamm.* **68**:860–864.

Busher, P. E., Warner, R. J., and Jenkins, S. H., 1983, Population density, colony composition, and local movements in two Sierra Nevadan beaver populations, *J. Mamm.* **64**:314–318.

Ermala, A., Helminen, M., and Lahti, S., 1989, Some aspects of the occurence, abundance and future of the Finnish beaver population, *Suomen Riista* **35:**108–118.

Gudkova, N. S., 1954, The effects of beaver reacclimatization in Gorki region and its impact on vegetation of water reservoirs' banks, Proceedings of III Ecological Conference, Part 3, Kiev.

Heidecke, D., Dombusch, M., 1990, Aktuelle Bestandsituation des Elbebibers in der DDR und methodische Anleitung zur Weiteren Bestandserfassung, Mitt. BAG Artenschutz, *Magdeburg* **13:**1–11.

Laptenok, V. V., Savchenko, A. P., 1997, Distribution and the numbers of the European beaver in the south of the Krasnoyarsk Territory, in: *Fauna and Ecology of Terrestrial Vertebrates of Siberia, Krasnoyarsk*, p. 156–168.

Milishnikov, A. N., Likhnova, O. P., Nikonova, O. A., 1994, Allozyme variability of European beavers, *Castor fiber* L. 1758, (Castoridae, Rodentia), *Russian J. Genetics* **30:**529–534.

Milishnikov, A.N., Saveljev, A. P., Likhnova O. P., 1997, Allozyme variation in European beaver (*Castor fiber* L., 1758) inhabiting the Berezina and Cheptsa rivers, *Russian J. Genetics* **33:**562–567.

Novak, M., 1987, Beaver, in: *Wildlife Furbearer Management and Conservation in North America*, (M. Novak, J. A. Baker, M. E. Obbard, and B. Malloch, eds.), pp. 283–312, Ashton-Potter Limited, Ontario Ministry of Natural Resources, Ontario.

Nummy, P., Pienmunne E., 1998, Riistatulokkaat menestyvat Polljolassa - mutta milla hinnalla? *Suomen Riista* **44:**18–29.

Safonov, V. G., 1962, Before the beaver harvesting, *Ohota i ohotnichie hozyaistvo* **2:**24–27.

Safonov, V. G., Semyenov, B. T., 1962, Ecological substantiation and organization forms of economic use of European beaver, in: Collected Papers of NTI VNIIZHP, No 4 (17), p.3–16, Kirov.

Zharkov, I. V.,1969, European beaver number dynamics in the sites of their reacclimatization, in: Proceedings of Voronezh State Preserve, Issue XVI, p.157–167, Voronezh.

5

BEAVER MANAGEMENT IN THE BALTIC STATES

Martins Balodis,[1] Nikolai Laanetu,[2] and Alius Ulevicus[3]

[1]State Forest Inventory Institute
30 Kristapa iela, Riga, LV-1046, Latvia
[2]Zoology Museum (BGZM), University Tartu
Vanemuise 46-313, 51014, Tartu, Estonia
[3]Institute of Ecology, Akademijos 2, 2600 Vilnius, Lithuania

ABSTRACT

Beavers have been an important component of the environment in the Baltic region for the past 10,000 years. Beavers were used intensively by humans for fur, meat, and medicinal products for thousands of years. This exploitation led to the almost complete extirpation of beavers from the Baltic region by the late 1800s. Beaver reintroduction programs were initiated in the early to mid 1900s, and the populations recovered and grew. The reintroduction programs were supplemented by the translocations of animals within the countries. In the 1960s (Lithuania) and in 1980 (Latvia) beaver hunting as a management tool began. Hunting was initially intense, but has slowed in the 1990s. The current population estimate for this region is 100,000 animals. There is also movement (immigration and emigration) of animals among the three countries and between the countries and Russia. The long-term role of beavers as a keystone species in aquatic habitats is considered critical in the region, and far outweighs any short-term economic damage they may cause.

1. INTRODUCTION

1.1. General

European beavers, *Castor fiber* L., the largest palearctic rodent, came into the Baltic region in the early Holocene, approximately 10,000 years ago. Humans were also present, although for approximately 90% of this time period the impact by humans on the environ-

Beaver Protection, Managment, and Utilization in Europe and North America, edited by Busher and Dzięciołowski.
Kluwer Academic / Plenum Publishers, New York, 1999.

ment, and beavers, was minimal. These 9000 years might truly be called the age of the beaver in Baltic region. Beaver population density during this time period most likely depended on local climate changes, hydrography, and predation intensity, and thus was determined by the carrying capacity of the environment for beavers. Soon after beavers settled all suitable habitats (rivers, lakes and watersheds), beavers functioned as a keystone species of the wetland ecosystems they inhabited. They greatly influenced the quarternary geology, and landscape formation, especially by the formation of the beautiful river valleys of the Baltic. This process of ecosystem modification has been best described for North American beavers by R. J. Naiman and his co-workers (Naiman and Melillo, 1984; Naiman, Melillo, and Hobbie, 1986; Naiman, Johnston, and Kelley, 1988; Naiman, Pinay, Johnston, and Pastor, 1994).

1.2. Territory and Early History

Three Baltic States (Latvia, Estonia, Lithuania) covering an area of 170,000 km^2, and with a human population of 8 million, are located at the east coast of the Baltic sea between 21°–28° east longitude and 54°–60° north latitude. The relief is moderate with the highest points about 300 m above sea level in Southeast Estonia and central Latvia. The annual precipitation is about 700 mm.

In addition to their considerable influence on the environment in the past, beavers also have had a great impact on human culture in prehistoric and historic time. Estonian paleontologist K. Paaver has discovered materials that date to the middle of the third millennium B.C. (about 5000 years ago). Paleontological excavations in Kaulenkalns and Zvejnieki have revealed that beaver bones represent 66% and 59% of all bones from animals found at these sites respectively. The meat, valuable fur, and especially the scent glands, which were used widely as medicinals, were the reasons for intensive beaver hunting. The popularity of beavers is testified also by toponyms, personal surnames, and folk songs (quatrains) of this region.

From this evidence we conclude that significant beaver hunting began in the third millennium B.C. and reached the highest intensity in the first millennium A.D. After this time, beaver hunting declined and beaver fur and meat products appear to have decreased in economical significance by the middle of the second millennium A.D. Evidence for a decline in the size of the population, which led to the need for legislated protection of beavers, is supplied by statutes regulating beaver management in Lithuania and Russia published in the 16th century. A decline in the societal importance of the animal may be inferred by the change in name of a street in the center of Riga, Latvia, from "Beaverstrate" to "Weverstrate" in the 15th century (Biernath, 1929). It may be that there was no cultural memory of the importance of beavers to the society by this time which led to the name change.

As a result of human exploitation there were virtually no beavers in the Baltic region by the 19th century. The last wild beaver was killed on the Rauna River in Latvia between 1871–1873. In Estonia the last beaver was killed in 1841. Beavers would be absent from Latvia for 56 years and from Estonia for 110 years. However, in Lithuania some beavers were still present in the 1930s, and it appears that they were never completely eliminated in this country. In these three countries the decline or total elimination of beavers allowed the aquatic ecosystems to develop without the influence of this vital biological component. Thus, the wetlands of this region were functioning in a different manner by the early 1900s than they had been centuries before when beavers were present.

1.3. Recovery of the Species—Reintroductions

Recovery of the species in Latvia began in August, 1927 when two mated pairs were released on the Stende River (Irbe River basin), which flows to the Baltic Sea. In 1935 another mated pair was reintroduced on Lipsa Creek at the location where the last wild beaver in Latvia was killed. All of the reintroduced beavers came from Norway, were brown in color, and were the subspecies *C. f. fiber.* By 1939 there were at least 60 beavers in these areas. In 1952 five additional pairs of black beavers were released in the river basins of the Venta, Irbe, and Slocene. These pairs quickly formed permanent settlements along the river systems. A total of 16 foreign animals were introduced.

Beaver reintroduction began in Estonia in 1957 when five pairs of Belorussian beavers were released. These animals were introduced into the river basins of northern Estonia.

In Lithuania reintroduction of foreign beavers began in 1947, when eight beavers from Voronezh, Russia, were introduced. In 1948, an additional 40 individuals were brought to Lithuania. These introductions were further supplemented in 1959 by an additional 30 animals. Beavers used in these introductions came from the Gomel region, in Belorussia. A total of 78 foreign animals were used in the Lithuanian reintroduction program.

1.4. Recovery of the Species—Translocations

In addition to the documented reintroductions, the recovery of the beaver populations in the Baltic countries has been supported by the active translocation of existing animals. In Latvia, 145 beavers were released in the river basins (except in the Salaca, Lielupe and Velikaya rivers), in Lithuania, eight beavers were released in the Lielupe River basin, and in Estonia, according to Nikolai Laanetu, 115 beavers were resettled in 20 places of continental Estonia. These beavers were captured in the southeastern part of Estonia and therefore were of Russian origin. In the formation of the present beaver populations in the Baltic countries the movement of animals between countries, and the immigration of animals from Russia has been significant. Immigration and emigration is particularly important between Latvia and Lithuania.

2. METHODS

It is difficult to accurately estimate the number of beavers in specific locations, and during specific seasons. Population estimates can change seasonally. For example, foresters conduct wildlife counts in early spring before the kits (young-of-the-year) are active. This leads to a lower population estimate than if a count was conducted in late summer when the presence of kits could be documented. Methods for counting depend on the size of the population, qualification of the workers, and the financial resources available. Thus our estimates are influenced by these constraints.

In Latvia, from the first years after the reintroductions up to about 1939, the beavers were monitored constantly by the State Forest Service. These data are particularly credible and therefore were used to calculate annual growth rates of approximately 25% during this period. In the following years during World War II and up to the 1950s, beavers in Latvian woodlands were not counted. Since the 1950s, there have been annual, registered beaver counts, carried out by the State Forest Service, and monitored periodically by scientists. The situation is similar for Lithuania and Estonia.

3. RESULTS

3.1. Latvia—A Case Study of Beaver Population Growth

The modern history of the beaver population in Latvia is separated into three stages of the development. The first stage is the initial recovery of the species between 1927 and 1960. In these 33 years (which include the hard years of World War II) the number of the animals reached approximately 100 animals. The growth rate for this period was quite low (about 10%). The second stage is a consolidation in Latvian fauna and includes the time from 1960 to the start of beaver hunting in 1980. The number of beavers in the population during this time was estimated to be in the thousands, and the growth rate for the period of consolidation was 23% annually. The third stage represents one of active beaver management, (1980 to the present). Balodis (1998), based on a study of beaver in Gauja National Park, found that the carrying capacity of the environment for beavers in this region of Latvia had been reached. He also reports that the annual growth rate is 10% which is 2.3 times lower than reported for the second stage. The best estimate of beaver numbers in Latvia at this time is 70,000 individuals (Balodis, 1998). The beaver population in Latvia is expected to continue to increase and there may be 100,000 individuals in the country by the year 2000.

3.2. Hunting as a Management Tool

Harvesting, by trapping or shooting, is commonly used as a tool to manage beaver populations (see the papers by Saveljev and Safonov, Harkonen, and Hartman in this volume). Under the supervision of the Russian beaver-specialist B. P. Borisov, beaver hunting in Latvia began during autumn 1980 in the Venta River basin. During the first season many beavers were taken that had tail wounds from bites. A high incidence of tail wounds is often observed in high density populations, and suggests that the number of beavers has exceeded the carrying capacity of the habitat. The harvest rates rose rapidly, and reached more than 5,000 beavers annually by the late 1980s. In these 10 years of intensive hunting about 40,000 beavers were trapped. The harvests collapsed in the early 1990s due to a collapse in the fur trade, but after 1993 the hunting pressure began to slowly increase.

Hunting was also instituted in both Lithuania and Estonia. In Lithuania the beaver harvest began in 1967, and by 1992 about 25,000 animals had been caught. In the beginning of the 1990s 2,000 to 3,000 beavers were taken annually (Mickus, 1995). In Estonia beaver hunting has been justified by the extensive damage beavers cause to various regions of the country. However, many of the damage complaints against beavers are found to be baseless, and many others are exaggerated (Laanetu, 1995). Thus, in Estonia, beaver hunting may not be justified.

4. DISCUSSION

The recovery of the beaver, which functions as a natural conservationist in the ecosystem, in the Baltic Countries has continued for the past 70 years. As a result the Baltic region currently has an increasing beaver population of not less than 100,000 animals.

For Latvia, habitat suitability was determined by examining present beaver density. The number of beaver settlements in 88 km^2 plots were determined using questionnaires conducted by the State Forest Service in 1996. They were then corrected by conducting field investigations from a boat, by walking marsh routes, and from traplines. These data were used for population interpolations and extrapolations. The heterogeneity of beaver distribution suggests a remarkable potential for population growth in most localities. The more effectively beavers interact with their environment the higher the carrying capacity of that environment should be. Unfortunately, we can not currently predict what the carrying capacity will be in the future.

Because of the current global decline in the value of fur, beavers will not be judged as an economic resource, but rather by their contribution to, or influence on, the environment. The interaction between beavers and humans is always a fine balance between the value of the ecosystem services beavers provide, and the economic cost of these services to humans. The costs of having beavers in the environment include, flooded land, destroyed agricultural crops, loss of valuable forest trees to cutting and flooding, and the flooding of roadways. Beavers, as a keystone species in aquatic habitats, provide at least three major ecosystem services: 1) Beavers act as regulators of water flow in both small and large streams. This activity can be critical in the hydrodynamics of the stream during extreme conditions of flooding and drought. 2) Beavers can intensify the process of aquatic self-purification. Their activity of constructing dams, which reduces water flow and increases the rate of sedimentation in the ponds they create, can act as purifying agent. 3) The activity of beavers in an area increases the biological diversity of the area. It is apparent that in developed countries where human activities most often function to reduce biological diversity, beavers function in the opposite manner and actually increase biological diversity. Thus, if humans are to move towards a model of reconciliation ecology with wildlife the protection and conservation of beavers is vital.

REFERENCES

Balodis, M., 1998, Beavers in the Gauja National Park, Latvia, *Proc. Latvian Acad. Sci., Section B*, **52**:49–53.

Biernath, H., 1929, *Die Alten Strassennamen von Riga*, Universitäts Verlag, Leipzig.

Laanetu, N., 1995, The status of European beaver (*Castor fiber* L.) in Estonia and its influence on habitats, in: *Proceedings of the Third Nordic Beaver Symposium*, (A. Ermala and S. Lahti, eds.), pp. 34–40, Finnish Game and Fisheries Research Institute, Helsinki.

Mickus, A., 1995, The European beaver (*Castor fiber* L.) in Lithuania, in: *Proceedings of the Third Nordic Beaver Symposium.*, (A. Ermala and S. Lahti, eds.), pp. 44–45, Finnish Game and Fisheries Research Institute, Helsinki.

Naiman, R. J., and Melillo, J. M., 1984, Nitrogen budget of a subarctic stream altered by beaver (*Castor canadensis*), *Oecologia* **62**:150–155.

Naiman, R. J., Melillo, J. M., and Hobbie, J. E., 1986, Ecosystem alteration of boreal forest streams by beaver (*Castor canadensis*), *Ecology* **67**:1254–1269.

Naiman, R. J., Johnston, C. A., and Kelley, J. C., 1988, Alteration of North American streams by beaver, *Bioscience* **38**:753–762.

Naiman, R. J., Pinay, G., Johnston, C. A., and Pastor, J., 1994, Beaver influences on the long-term biogeochemica characteristics of boreal forest drainage networks, *Ecology* **75**:905–921.

6

THE REINTRODUCTION OF EUROPEAN BEAVER, *CASTOR FIBER*, IN POLAND

A Success Story

Ryszard Dzięciołowski[1] and Jan Gozdziewski[2]

[1]Department of Forest Zoology and Wildlife Management
Agricultural University SGGW
Rakowiecka 26/30, 02-528 Warsaw, Poland
[2]Provincial Board of Polish Hunting Association
Suwalki, Poland

ABSTRACT

The European beaver, *Castor fiber*, a species native to Poland, was completely extirpated by 1945. After this date a number of beavers immigrated to Poland from Lithuania, and others were actively introduced from the former Soviet Union. The population grew slowly from 1945 to the 1970s. In the late 1970s the population began to increase rapidly, and by 1997 there were an estimated 10,000 beavers in Poland. The growth of the population was influenced by the active protection of beavers, and by the relocation of animals into suitable habitats. The success of the reintroduction, protection, and relocation programs has led to a large, viable beaver population in Poland. It is now necessary to implement an active management program for this species, a component of which would be an annual harvest of beavers. An annual harvest will reduce damage problems due to beaver activity, and help insure the continued health of the Polish beaver population.

1. INTRODUCTION

The European beaver, *Castor fiber*, was native to Poland. In the middle ages beavers were found in all areas of Polish territories, except in the high mountains. Through the centuries the range of Polish beavers was reduced until it reached a minimum in the 19th century. At the beginning of 20th century the beaver population was very small, and restricted

Beaver Protection, Managment, and Utilization in Europe and North America, edited by Busher and Dzięciołowski.
Kluwer Academic / Plenum Publishers, New York, 1999.

to the eastern part of the country. At this time attempts had already begun to restore *C. fiber* in Poland. World War II, and a resulting shift of national boundaries, caused the complete extirpation of beavers from Poland by 1945. In the following years several individuals migrated from Lithuania to the Neman watershed in Poland. The first reintroductions in Poland were successfully carried out on the Biebrza River in 1948 and 1949. The beavers used in this reintroduction program were brought from Voronezh in the former Soviet Union.

2. THE GROWTH AND EXPANSION OF THE BEAVER POPULATION

Beavers from these two sources (Lithuania and the Soviet Union) were the founders of the present Polish beaver population (Table 1). It is evident from these data that the beaver population grew slowly through the mid 1970s. In the period from 1945 to 1972 the population increased from 0 to 350 animals. By the late 1970s the population growth rate accelerated rapidly. In the 20 year period from 1977 to 1997 there was an increase of 9000 animals in the total population and an average annual increase of 450 animals.

3. THE BEAVER PROTECTION AND RELOCATION PROGRAM

3.1. The Management Program

The acceleration in population growth rate was influenced by the implementation of a program of active protection and relocation of beavers which was initiated in 1974. The program was developed by scientists from the Polish Academy of Sciences (and personally by the late Professor W. Zurowski), and carried out in cooperation with the Polish Hunting Association. The main objective of the program was to move beavers from the Neman River and Pregola River watersheds to the watersheds of the Vistula and Odra rivers. Individuals were captured in areas densely populated by beavers in northeastern Poland and were released in the southern and western parts of the country. The goal was to create a continuous beaver population in these regions.

3.2. Capture and Relocation Methods

Capturing beavers is accomplished by surrounding lodges or dens with nets and flushing the animals out with dogs. A special breed of small, non-aggressive, but persistent, dog is used for this purpose. Beavers are captured in spring (from the time the ice melts through 20 April), and in autumn (October). On average 4.5 beavers are captured during a day at a cost of US $60.00 per animal. Families (adults, yearlings and kits—the

Table 1. Estimated number of European beavers in Poland (1945–1997)

Year	Estimated number	Year	Estimated number
1945	probably none	1977	1000
1958	130	1987	2900
1966	270	1994	7000
1968	300	1997	10,000
1972	350		

young of the year), when captured together, are released together in a new location. Two-year-old and three-year-old individuals (who have not formed a pair bond at the time of capture) are used to form pairs with animals from different populations. Beavers are transported by car (this does not take more than 14 hours), and casualties during captures and transportation do not exceed 3%.

In selected locations at least 4 families are released to form the nucleus of a new population. These newly colonized sites are visited in the following years, the population is observed and censused, and any losses are supplemented with new individuals. Local hunting organizations look after the release sites and protect beavers against poaching and vandalism.

3.3. Program Results

The program has been in effect for the past 25 years, and relocation activity has intensified since 1995 (Table 2). The results are also presented graphically in the two figures. Figure 1 illustrates the distribution of beavers in Poland as of 1937, while Figure 2 shows the distribution as of 1997. Comparison of these two maps illustrates the rate of reintroduction and formation of new beaver colonies in southern, central and western Poland.

Beavers have revealed a remarkable potential for adaptation to unnatural, and often adverse, conditions (such as water pollution and human disturbance). There is evidence that beavers may even change their normal nocturnal pattern of activity and become more active during the day. Beavers have also been able to withstand periodic, and substantial, drops in water level in these new sites.

Mortality during dispersal after release is higher in the plateau and mountain sites than in lowland areas. In the higher elevation areas mortality may be as great as 35%. The period of spring thaw, and the resulting dispersal of animals, is the critical time for beavers. Spring migrations may be as far as 50 km. There are also indications that wolves, which are efficient beaver predators, may prevent colonization of certain areas by beavers.

4. BEAVER MANAGEMENT IN THE FUTURE

As a result of reintroduction and relocation activities Poland has an expanding beaver population, free of parasites and diseases, which covers almost all suitable habitats in the country. Beavers have traditionally been a protected species in Poland. However, the

Table 2. Number of beavers released in various regions of Poland during successive years

Year	Number of animals released	Province
1988	10	Rzeszów
1989	21	Olsztyn, Toruñ, and Kielce
1990	20	Piotrków and Sùupsk
1991	24	Rzeszów, Koszalin, and Czech Republic
1992	40	Rzeszów, Cheùm, and Czech Republic
1993	37	Bielsko-Biaùa, Przemyúl, and Kraków
1994	46	Przemyúl, Lublin, Legnica, and Krosno
1995	86	Wùocùawek, Rzeszów, Krosno, Nowy Sàcz, and Sùupsk
1996	131	Bielsko-Biaùa, Cheùm, Krosno, Lublin, Rzeszów, and Tarnobrzeg
1997	120	Bielsko-Biaùa, Przemyúl, Krosno, Rzeszów, Sùupsk, Waùbrzych, and Opole
Total	535	

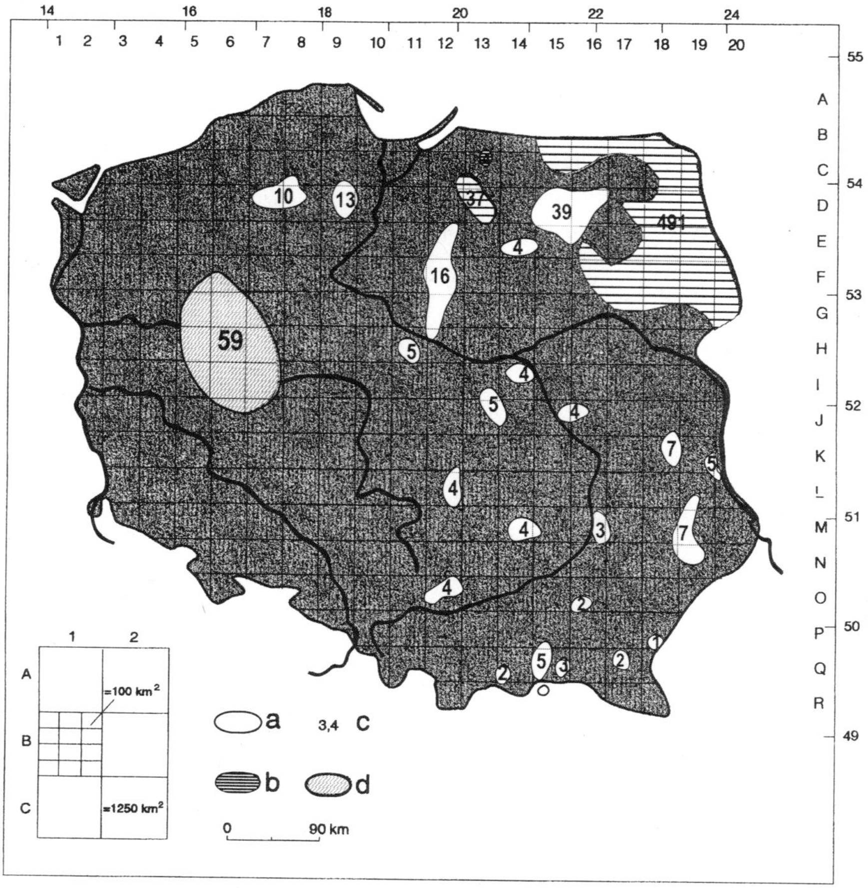

a - population reintroduced in 1975-1986 in the Vistula River watershed
b - resident beaver populations came into existence in result of reintroduction carried out in forties as well as of immigration
c - number of beaver sites in the population
d - population introduced into Odra River watershed in 1974-1983 (compiled on the data: Graczyk 1985)

Figure 1. The distribution and numbers of the European beaver population in Poland in 1986–1987.

current status of the population suggests that it is time to investigate instituting a beaver harvest in Poland. The beaver population in northeastern Poland reveals symptoms associated with high density. Beaver damage to farm land and agricultural crops, and to forests is increasing. The government, which is paying compensation to farmers for beaver damage, wants to declare the beaver a game species in the hope that hunting will alleviate the current damage problems. However, Polish hunters currently are reluctant to shoot beavers, and the practice of trapping them is completely forgotten. The problem with using trapping to help manage the beaver population is that trapping was never popular in Poland, and is considered the procedure of poachers. Thus, trapping is not viewed as a legal activity. For an annual beaver harvest to be an effective management tool both of these issues (lack of desire to shoot beavers, and no trapping history) will have to be overcome.

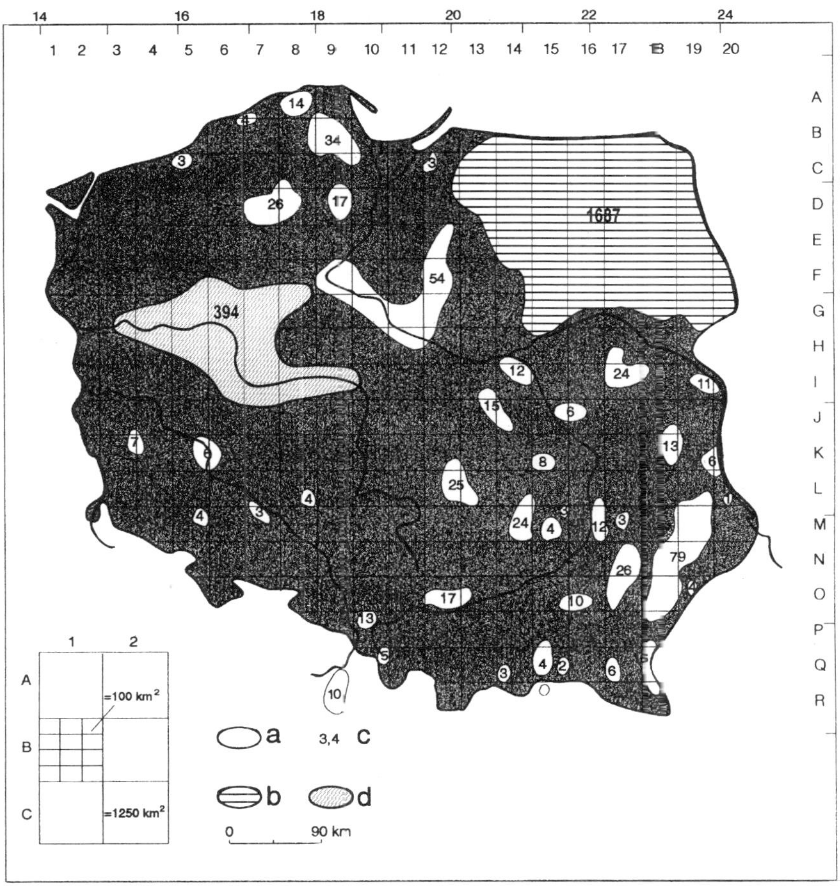

a - population reintroduced in 1975-97 (Żurowski and Goździewski)
b - population formed in 40-ties through imigration
c - number of families within population
d - population reintroduced in the Odra River watershed in 1974-1983

Figure 2. The distribution and numbers of the European beaver population in Poland in 1996–1997

The growth of the beaver population in Poland has reached the point where firm decisions about beaver management are required. The implementation of a scientifically sound management program, which would include a beaver harvest, will help insure the future health of the Polish beaver population.

7

THE AUSTRIAN BEAVER, *CASTOR FIBER*, REINTRODUCTION PROGRAM

Johanna Sieber

Konrad Lorenz Institute for Comparative Ethology
Austrian Academy of Sciences
Savoyenstrasse la A-1160 Vienna, Austria

ABSTRACT

The European beaver, *Castor fiber*, was native to Austria. However, beavers were completely extirpated from Austria in the 1860s. In the 1970s and 1980s beavers were reintroduced to Austria along the Danube River east of Vienna. This east Austrian population has grown and spread. Smaller populations also exist in northern and western Austria. Current population estimates are approximately 800 animals in the eastern population and 200 animals in the western population. There are also 15 smaller groups scattered in other parts of the country. The growth in numbers and geographic spread of beavers in Austria have produced some conflict between beavers and humans. Scientifically sound management is necessary to ensure the conservation of beavers in Austria.

1. INTRODUCTION

Nearly all European beaver (*Castor fiber*) populations with the exception of small groups in France, the former East Germany, Norway, and possibly Poland, were destroyed by over-hunting during the last century (Djoshkin and Safonov, 1972;, Hartman, 1995; Dzieciolowski and Gozdziewski, this volume). In the early 1900s Sweden and Finland began to reintroduce European beavers by releasing beavers from Norway. Following this model, some countries in central Europe (Switzerland and the former West Germany) started reintroductions in the 1960s and 1970s (Jacob, 1998; Rahm and Baettig, 1996). At the same time the remnant populations recovered under complete protection, the reintroduced beaver families built up viable populations in the above mentioned countries (Heidecke, 1991; Nolet and Rosell, 1998; Zurowski and Kasperczyk, 1988). The Austrian beaver population was already overexploited 200 years ago; beavers then were rare and

Beaver Protection, Managment, and Utilization in Europe and North America, edited by Busher and Dzięciołowski.
Kluwer Academic / Plenum Publishers, New York, 1999.

highly conspicuous game animals hunted for fur, meat and musk. On Austrian territory the last beaver was killed in 1863 in the Danube estuaries east of Vienna (von Mojsvar, 1897).

2. BEAVER REINTRODUCTION AND DISPERSAL

More than a hundred years after the last Austrian beaver was killed, and after becoming aware of the successful beaver reintroductions in the former West Germany and Switzerland, a decision was made by the late Otto Koenig (former director of the Konrad Lorenz Institute) to reintroduce beavers to Austria (Koenig and Krebs, 1979). Habitat suitability was investigated and the best sites for the first beaver reintroductions were found in the Danube estuaries between Vienna and the Slovakian border (Figure 1). This area has more than a hundred kilometers of slow running water, or still waters adjacent to the river on both sides, and enough softwood and other vegetation to sustain a growing beaver population. Preference for introduction was given to the eastern subspecies, *Castor fiber osteuropaeus*, and *C. fiber belorusicus* for two reasons: (1) it was easier for us to purchase them (mainly from Poland and the former USSR), and (2) the Danube water system is more closely connected to the eastern river systems than it is to western rivers.

Between 1976 and 1982 approximately 40 beavers were released. The introduced beavers were mainly *C. f. osteuropaeus, C. f. belorusicus*, and Polish hybrids, but some *C. f. fiber*, and unfortunately a few *C. canadensis* were also released (Sieber, 1995 and 1998). Austrian reintroduction efforts began slowly and a number of problems were encountered. Some of the beavers were found dead (for unknown reasons these were primarily North

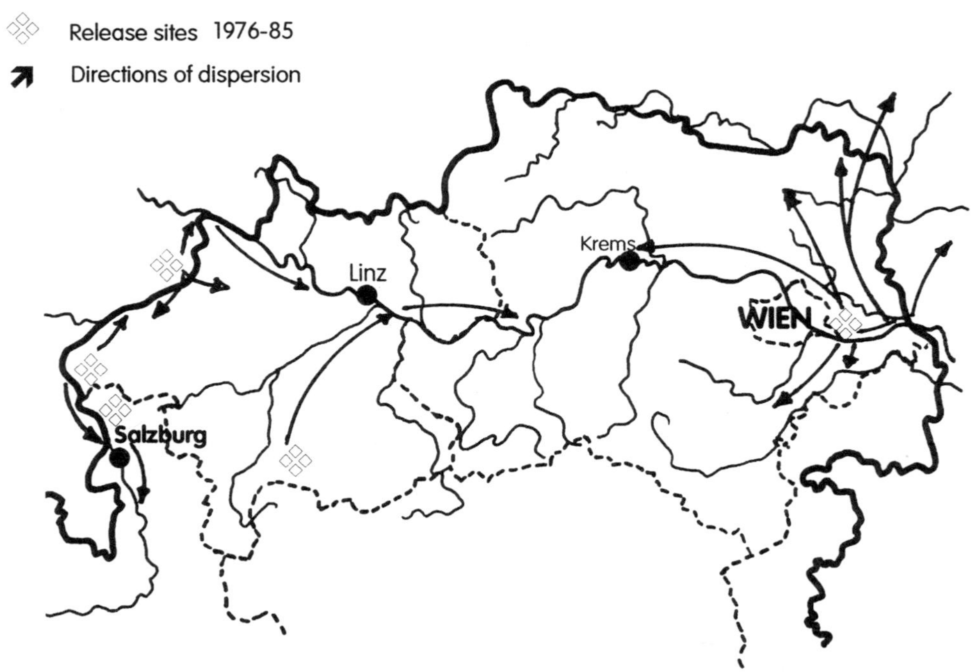

Figure 1. Map showing the release sites of beavers in Austria, 1976–1985. The initial release site was near Wien (Vienna). Arrows indicate the dispersal patterns of the released beavers.

American and Russian animals), some of the beavers disappeared, and only a few of the introduced pairs reproduced successfully. Therefore, an additional 15 animals were released later. The population did not show any increase until 1985 (Kollar and Seiter, 1990). Beginning in the late 1980s the first animals migrated from the left bank to the right bank of the Danube and invaded smaller tributaries establishing new territories. The sites of all reintroductions and dispersal patterns are presented in Figure 1.

In the late 1980s beavers also began to disperse to the river March (Morava), which forms the border between Austria and Slovakia. By 1988 the first individuals had settled in the Slovakian tributaries of the river March. The present Slovakian population in the Zahorska plains, which includes more than 20 beaver colonies (an estimated 80 individuals), originated from the Austrian reintroduction project (Pachinger, 1994; also see the paper by Pachinger and Hulik in this volume). Three years ago the dispersion continued to the Czech river Dije (Thaya) which borders Austria in the Northeast. Some beavers also dispersed along the Danube to Hungary. The results of a beaver census in 1988–1989 found continuing dispersal. However, the total east Austrian population was estimated to be only 100–150 individuals (Kollar and Seiter, 1990). The number of North American beavers in the population was unknown at that time.

Near the Austrian capital of Vienna, dispersing subadults were observed both north and south of the city and, most surprisingly, some even found their way straight through the city traveling more than 30 km along artificial channels (Donau-Kanal) and a completely channeled creek (Wienfluss). During the last 5 years some beaver settlements have been established within the Vienna city limits, primarily in suboptimal habitat. However, highly adaptable animals like beavers are able not only to survive, but to colonize and reproduce in these marginal habitats. (Sieber, in prep.).

In Upper Austria in 1985, after escaping from the enclosures of a private animal park, Swedish beavers (*C. f. fiber*) founded a small but rapidly growing population. From this location the subadults dispersed to the north and settled at the river Traun (leading to the Danube), and soon after in the wetlands on both sides of the Danube itself (Sieber, 1989). Meanwhile an additional west Austrian population became established. This population (mainly the subspecies *C. f. fiber*) originated from animals released on the Bavarian banks of the Inn River between 1970 and 1974. This group experienced the same problems as the eastern population (high mortality and slow growth rate) in the first few years after reintroduction, but soon began to flourish as the dispersing subadults discovered some excellent habitats on the Austrian side of the bordering river (Reichholf, 1976). Dispersing individuals moved more than 100 km downstream and there are presently beaver colonies between Passau and Linz (Sieber, 1989). Currently, dispersing beavers from the eastern population have already reached these western settlements along the Danube.

3. CURRENT STATUS OF THE AUSTRIAN BEAVER POPULATION

After a survey of the eastern Austrian beaver population in 1997/98 I estimated a total of at least 200 settlements (Sieber, 1998), and approximately 800 individuals (following the international European convention from Heidecke, 1984). This estimate did not include either Slovakian or Czech beaver sites. The west Austrian population was surveyed last in 1994–95 and approximately 50 family groups were found (Figure 2). An additional 15 scattered groups live in different parts of the country.

Unfortunately, beavers have quickly dispersed from public lands where they were released to privately owned, agriculturally valuable land. As damage to land, trees, and crops

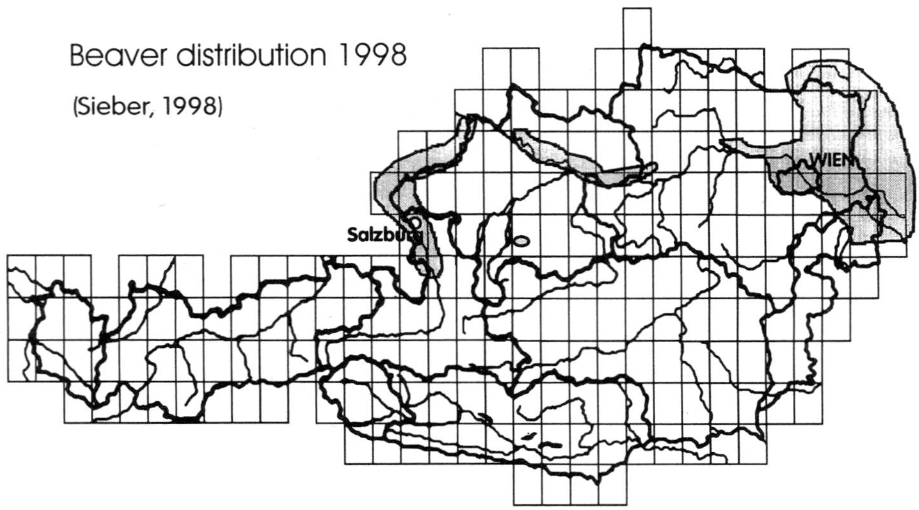

Figure 2. Map showing the current (1998) distribution of beavers in Austria. The shaded areas represent the approximate ranges of the three Austrian sub-populations.

caused by beaver activity increased, public opinion about beaver reintroduction has changed. What began as enthusiastic support for the beaver program changed gradually to relative indifference. Currently, there is even some open opposition to the reintroduction program. All of those involved in the beaver reintroduction program must now develop management procedures which are not in conflict with the aims of the original beaver conservation plan.

4. MANAGEMENT RECOMMENDATIONS

A careful review and analysis of the status of beavers in Austria allows me to present the following recommendations. These recommendations are designed to reconcile beaver protection and conservation with effective beaver management.

1. Survey the populations at least every 3 years, document the distribution of beavers, and attempt to predict the extent of future beaver colonization.
2. Attempt to predict "sensitive areas" where beaver damage is to be expected and beaver-human conflicts will occur.
3. Identify locations where beavers are to be tolerated, and those where they will not be tolerated. While these may be primarily socio-economic and political decisions, they should be made following the recommendations of conservation biologists.
4. Enlarge the existing nature conservation areas and set aside river bank habitats by excluding forestry and agricultural production in these areas. This will allow these human-modified habitats to return to their natural condition.
5. Develop a "damage compensation policy" for beaver that is similar to the policies established for other protected species such as otter and bear. The compensation policy should be closely dependent on human tolerance of future damage.
6. Provide scientific and management advice, and money for protective measures like fences, repellents, and dam drainage systems.

7. Livetrap and relocate "problem animals".
8. Consider initiating hunting regulations which will allow sustainable use of the beaver population in the future.

REFERENCES

Djoshkin, W. W., and Safonov, V. G., 1972, *Die Biber der Alten und Neuen Welt. Die Neue Brehm Bücherel* **437**, A. Ziemsen Verlag Wittenberg-Lutherstadt.

Hartman, G., 1995, Patterns of spread of a reintroduced beaver population in Sweden, *Wildlife Biology* **1**:97–103.

Heidecke, D., 1984, DDR-Übersicht über die Bieberbestandserfassung im Jahr 1982 und organisatorische Hinweise, Informationan zum

Biberschutz der Biol. Station Steckby des ILN Halle 1983/2, Mitt. BAG Artenschutz *Magdenburg* **6/5**, Dez, 83.

Heidecke, D., 1991, Zum Status des Elbebibers sowie etho-ökologische Aspekte. *Seevögel* **12**:33–38.

Jacob, J. C., 1998, Le Castor dans le sudest de la France. *Office National de la Chasse Paris.*

Koenig, 0., and Krebs, U., 1979, *Bedeutung und Methodik der Ansiedlung von Bibern*, Forschungsgemeinschaft Wilhelminenberg Wien.

Kollar, H. P., and Seiter, M., 1990, *Biber in den Donau-Auen östlich von Wien. Eine erfolgreiche Wiederansiedlung, Umwelt* **14**. Verein für Ökologie und Umweltforschung, Wien.

Nolet, B. A., and Rosell, F., 1998, Comeback of the beaver *Castor fiber*: An overview of old and new conservation problems, *Biological Conservation* **83**:165–173.

von Mojsvar, A. M., 1897, *Das Thierleben der Österreichisch-Ungarischen Tiefebene, Biologische und thiergeographische Skizzen und Bilder*, Alfred Hölder, Wien.

Pachinger, K., 1994, Spontane Verbreitung des Bibers im Flußtal der March und der Zahorska Tiefebene. *Ekologia Bratislava* **1994**:Suppl. 1.

Rahm, U., and Baettig, M., 1996, Der Biber in der Schweiz. Bestand, Gefährdung, Schutz, *Schriftenreihe Umwelt* **249**. Hrsg. Bundesamt für Umwelt, Wald und Landschaft.

Reichholf, J., 1976, Die Ausbreitung eingesetzter Biber am unteren Inn, *Mitt. Zool. Ges. Braunau* **2**:361–368.

Sieber, J., 1989, Biber in Oberösterreich, Eine aktuelle Bestandsaufnahme an Inn und Salzach, *Wiss. Mitt. OÖ Landesmus*. **7**:397–405.

Sieber, J., 1995, Sie schwimmen wieder ! Biber (*Castor fiber*) in Österreich, *Stapfia* **37**:217–224.

Sieber, J., 1998, 20 Jahre nach der Wiederansiedlung. Biber-Management unvermeidlich? *Naturschutz und Landschaftspflege in Brandenburg* **1**:79–80.

Zurowski, W. and Kasperczyk G., 1988, Effects of the reintroduction of European Beaver in the lowlands of the Vistula Basin. *Acta Theriol.* **33/24**:325–338.

8

ORIGIN, PRESENT CONDITIONS, AND FUTURE PROSPECTS OF THE SLOVAKIAN BEAVER POPULATION

Karol Pachinger[1] and Tomas Hulik[2]

[1]Institute of Ecology
[2]Department of Ecosozology and Physiotactics
Comenius University
Mlynska dolina B-2, 842 15 Bratislava, Slovakia

ABSTRACT

The reintroduction of beavers into countries surrounding Slovakia (Austria and Poland) allowed the immigration of animals to Slovakia by the 1980s. Beavers are currently found in three main regions of Slovakia: 1) the Zahorska lowland, 2) the Danube River and floodplain near Bratislava, and 3) the Nizke Beskydy Mountains in northeastern Slovakia. Beavers were first observed in the Zahorska lowland in 1977, and there are currently at least 16 active beaver territories known in this region. Beavers have also dispersed to the Danube floodplain forests north and south of Bratislava where as many as 12 different areas have been occupied, or are currently occupied. Beavers immigrated into the Nizke Beskydy Mountains from Poland in the 1980s, and a small population is also found here. The Slovakian beaver population continues to increase in number, and expand in range. Limits to the range expansion of the beaver in Slovakia include human developments. For example, on the Danube River, downstream from Bratislava, the new Gabcikovo Dam limits the ability of beavers to disperse in this direction. We suggest a bypass be constructed near Cunovo to facilitate beaver movement.

1. INTRODUCTION

The first two European beavers, *Castor fiber*, were released to the wild in a floodplain forest near Eckartsau, Austria in 1976. An additional four animals were released in 1977. Shortly thereafter, more beavers were released in the Danube floodplain forest re-

Beaver Protection, Managment, and Utilization in Europe and North America, edited by Busher and Dzięciołowski.
Kluwer Academic / Plenum Publishers, New York, 1999.

gion of Austria. Between 1976 and 1988 a total of 42 beavers were released east of Vienna, Austria. However, 15 of these animals were North American beavers, *Castor canadensis.* Not all of the beavers stayed in Austrian territory. It was observed that they rapidly spread to the aquatic habitats in southwestern Slovakia. Beavers have also begun to occupy the northeastern region of Slovakia on the border with Poland. The first discovery in this region was made in 1981 when a bus hit a young male beaver. In 1987 three beaver settlements were found, two of which are inhabited at the present time. The areas currently occupied by beavers in Slovakia are shown in Figure 1. Thus, Slovakia is the only country in Europe with an expanding beaver population which is the direct result of beaver reintroductions in neighboring countries.

2. METHODS

2.1. Data Collection

Our knowledge about the presence and activity of beavers in Slovakia comes from our observations, or from the observations of a net of voluntary co-workers.

2.2. Regions Occupied by Beavers

The beavers were most successful in the Zahorska lowland in Slovakian territory (Figure 1). This region is in southwestern Slovakia, bordered by the Morava River on the south, by the Low Carpatian Mountains on the east, and by the Myjava Hills on the north. The region is rich in aquatic habitats, with many streams flowing from the western slopes of the Low Carpatian Mountains, and with many drainage canals. Many water basins were built here for the purpose of fish breeding, or they arose through gravel and peat production. Most of the basins have not provided suitable habitat for beavers. The whole region of the Zahorska lowland is extremely affected by human activity. Apparently, the most appropriate habitats for beavers, are the dead arms of the Morava River. The average annual temperature in the Zahorska lowland is 9°C (Luknis and Plesnik, 1961). The average annual maximum snow cover is between 15–20 cm, and the snow cover usually does not last longer than six months.

The second region, which is open to beaver immigration from Austrian reintroductions, includes the arms of the Danube River from Devin to Komarno (Figure 2). Although the best sites for beaver settlement in this area are the high earth-sand banks of the Danube arms which have ample amounts of willow and poplar, these sites are not utilized. The average temperature here is higher than in the Zahorska lowland. It is the warmest region of Slovakia, but the snow cover has the same character as in the Zahorska lowland. Although the Danube is a big European river which opens onto a huge plain and then reforms in new ways after passing through the Low Carpatian Mountains near Devin, it still behaves in many ways like a smaller mountain river. This is because the main source of its water comes from the Alps. Like mountain rivers, the Danube is subject to flooding from the end of April until the end of June, during which time a large amount of the snow in the Alps melts and fills the river.

The third region where a small number of beavers have settled is in the Nizke Beskydy Mountains in northern Slovakia (Figure 1). This is a low mountain range with an average elevation of approximately 600 m. Only a few of the summits reach the height of 1000 m. The climate here is relatively warm, but it has a continental character much like

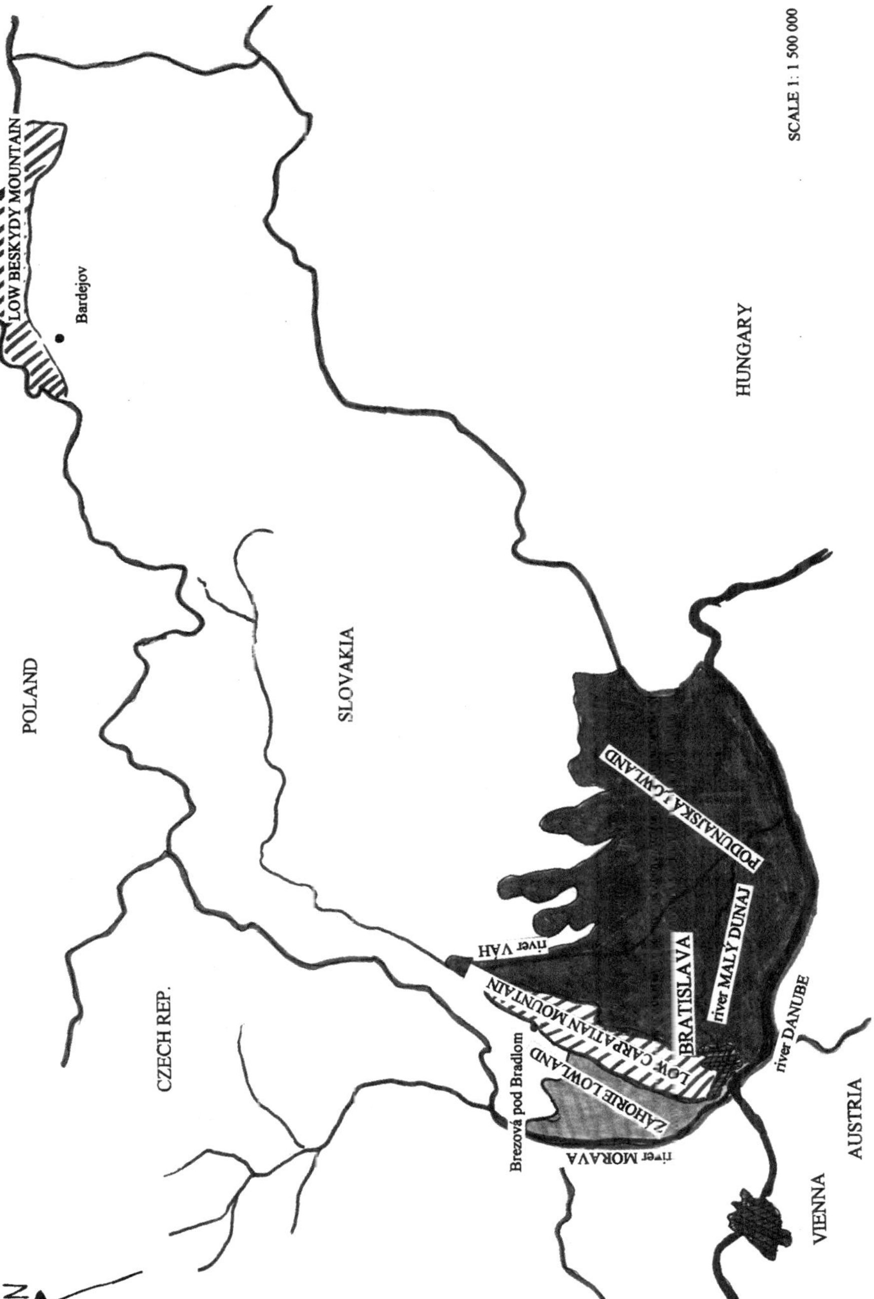

Figure 1. Map of Slovakia showing the current locations of beavers.

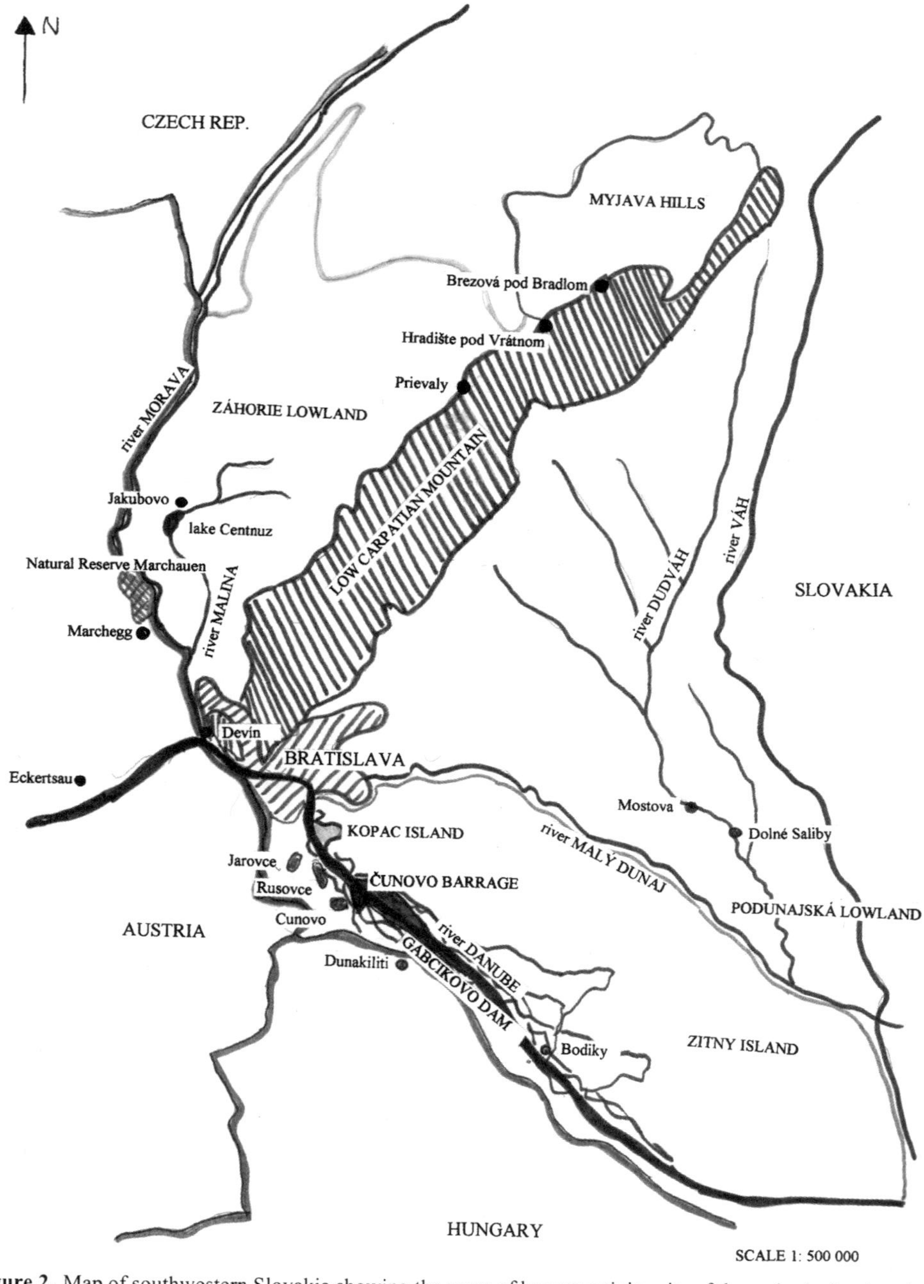

Figure 2. Map of southwestern Slovakia showing the areas of beaver activity, site of the major hydroelectric dam, and the water control system.

the Zahorska lowland. The average temperatures range between -3.6°C and -6°C in January, and between 14°C and 17°C in July. The region receives between 650 mm and 800 mm of rainfall annually, and is primarily a beech forest.

3. RESULTS

3.1. Zahorska Lowland

The first beaver was seen on Slovak territory after the initial attempts to reintroduce beavers near Eckartsau, Austria. Kollar and Seiter (1990) describe the exact chronology of releasing beavers into the Danube floodplain forest. Eckartsau, Austria is located only 15 km north of the confluence of the rivers Danube and Morava (Figure 2).

In July, 1977, a beaver was shot in the Zahorska lowland near the village of Jakubovo. It was the first beaver killed (the beaver was shot by accident) in Slovakia in 150 years. Today, the dermoplastic preparation of this beaver is kept at the Institute of Ecology of the Faculty of Natural Sciences of the Comenius University in Bratislava. The beaver was probably from the first group of animals that was released in Austria. If this is where the animal originated it had to travel down the Danube, up the Morava River, and then through the regulated Malina River. This is a 100 km long journey in a very short period of time. This event occurred during the period of the "Iron Curtain" when the arms of the Morava River were hermetically closed since they created the border between Czechoslovakia in the Soviet block and Austria in the west. Due to this political closure of the river, there is no available information on whether any additional beavers appeared there at that time. Only after a 10 year "vacuum" between 1977 and 1987 were beavers observed on a transverse-lake of Centnuz. Beavers were also found on spacious lakes that originated not far from the place of the first accidentally shot beaver near Jakubovo.

At the same time, in the winter of 1987/88, inhabited areas were observed on the Austria side of the Morava River near Marchegg (Natural reserve Marchauen), and also near the village of Stillfried. Stillfried is located on the other side of the Morava River, opposite Lake Centnuz near Jakubovo. Beaver settlements have been found more frequently in the Zahorska lowland since that time. Between 1988 and 1994 there were mainly transient habitations of young beavers apparently looking for appropriate sites for long-term settlement (Pachinger, 1994). Cuttings and transient habitations on banks were located on all streams of the Zahorska lowland. Accurate documentation is very complicated, but evidence of beaver activity suggests that many beavers passed through the region looking for the most optimal habitats. The first permanent beaver habitation (dam and lodge) was discovered in 1995 in Brezova pod Bradlorn, and on the same stream in Hradiste pod Vratnom village (Figure 2). The transient habitation of beavers was also observed in the summer of 1996 on the Dudvah River near village Mostova, and in the autumn on a lake near village of Dolne Saliby. In 1998 four other beaver dams and three beaver lodges were found. Currently there are 16 beaver territories known on the Zahorska lowland, and 10 - 12 of them are recorded on the Morava River and its arms. Not all of the sites are occupied by reproducing pairs. Therefore, some of these sites may be abandoned in a very short time.

3.2. Danube Floodplain Forests

According to Chernel (1886) the last beaver was caught on Zitny Island in 1858 (Figure 2). After that time beavers were considered extinct in Slovakia. Soon after the reintroductions in Austria, beavers were observed on the Slovakian section of the Danube. Permanent beaver occupation occurred earlier here than in the Zahorska lowland. Somogyi (1985) wrote in the magazine Ziva that two young beavers were living in the arm-system near Jarovce just outside Bratislava (Figure 2) from February 1984 until April 1985. A spring flood was the reason they left the location. However, in autumn 1985 and during 1986 beaver cuttings were found in the river arm system near Cunovo village, which is approximately 10 km downstream from the first location (Figure 2). After the unexpected flood in the summer of 1987, a beaver was caught in a fishing net by fishermen in Hungary near the village of Dunakiliti which is 8–9 km from Cunovo village (Stollmann and Voskar, 1989) From that time on transient beaver habitations were seen on the Danube arms, but long-term habitation was inhibited by flooding. The last beaver penetrated the river arm system and permanently settled down on the Sulianske River arm in autumn 1992.

Pachinger (1995) describes the exact chronology of beaver activity on the Danube River arms. In the spring of 1993 a water control barrier was built near the village of Cunovo, which prohibited beavers from moving freely along this section of the Danube. Beavers were not seen in this area until the winter of 1997/98. However, beaver lodges on the Rusovce River arm system above the Cunovo barrier started to increase. In the winter of 1993/94 the beavers settled at a formerly inhabited place on the Jarovce River arm, and on the left side of the Danube River arm which surrounds Kopac Island. These beavers were killed by poachers in 1994, but two other territories were discovered in 1995 on the Rusovce River arms near the Cunovo barrier. In 1996 three more sites were discovered. In 1997 eight inhabited sites were discovered in a relatively small area of the Rusovce River arms and the left river arm of Kopac Island.

For the third time the beavers settled on the Jarovce River arm on the outskirts of Bratislava. In the winter of 1997/98 two newly inhabited areas were discovered on this river arm system. One site is near the river kilometer-mark, #1927, which is near the middle of this length of the river arm system. The other site is near the river kilometer-mark, #1812, which is near the mouth of the river arm system of the old Danube River bed.

3.3. Northeastern Slovakia

The beavers found in northeastern Slovakia came from Bieszczad niskich, Poland. Beavers have been reintroduced to this area in Poland from the overcrowded Mazorskej lake plain since the beginning of 1980s by Polish scientists (Pucek, 1983, Tarnawski, 1987). According to Budajova (1995), only two out of the three inhabited areas discovered by Stollmann and Voskar in 1987 are still occupied. The number of beavers that live in these two locations is six (Budajova, 1995). It is interesting to note that since 1987 the areas have not been inhabited continuously but only temporally. It is also interesting that the beavers had to cross a ridge of the Beskydy Mountains on their journey from Poland to Slovakia, in order to settle on the South side of Beskydy Mountains.

4. DISCUSSION

4.1. Zahorska Lowland

Although there is no historical documentation of beaver presence in the Zahorska lowland, there is no doubt that the entire region of the Morava River was inhabited by beavers in the past. The region was rich in suitable aquatic habitat created by the tributaries of the meandering Morava River, and by streams flowing from the Low Carpatian Mountains. It is possible that beaver cuttings seen on Zitny Island, at a time when beavers had not lived there for at least 30 years, came from beavers living in a hidden location in the inaccessible swamps associated with the Morava River. Beavers could have easily migrated to the Danube, and then to Zitny Island. The first human water regulation structures on the Morava River were constructed at the turn of this century. Presently we are witnessing the rapid human resettlement of this territory. Although the whole country is affected by human activities, this land is especially well utilized. While wetlands are drained and worked intensively, beavers are still able to use the network of canals to move deeper into the land. Young beavers pass through the region when dispersing while looking for places for a permanent settlement. The most optimal habitats (Centnuz and Prievaly) were occupied first and reproduction was observed a year after the initial occupation.

The systematic manner in which the optimal sites were occupied is very interesting. It appears that migrating beavers passed through the whole region and chose the most optimal of all the suitable sites. Also, suboptimal sites were occupied, and we recorded a tendency for the dispersing beavers to penetrate deeper into the country, first to the north, and then to the southeast. In the region of Brezova pod Bradlom the ridge of the Low Carpatian Mountains is very low. This allowed beavers to cross approximately 1600 m over land to reach streams on the southern side of the Low Carpatian Mountains. This tendency for beavers to spread out and move relatively large distances has also been observed in northeastern Slovakia. Movement of dispersing animals is the only explanation for the population of beavers discovered on the Dudvah River, and at Dolne Saliby village. It is also possible that the beavers discovered in the winter of 1997/98 on the Danube tributaries near Palkovicovo and Bodiky villages, passed around the Low Carpatian Mountains, down the Vah River, and then up the Danube to the tributaries. It is remarkable that during the first year of looking for suitable sites beavers preferred proceeding deeper into the country instead of inhabiting the river arm systems of the Morava River. The best habitats along this river have only been settled in the last few years. One reason may be because the Morava River has annual fluctuations in water level of 2–3 m. No reproducing pairs have been discovered here yet. Until 1992 reproduction was observed only in the first two optimal habitats (lake Centnuz and Prievaly village), yet the number of migrating individuals was higher than expected if they were only dispersing offspring from the existing pairs. Therefore, we assume that beavers moved to our area from Austria where they were actively being introduced. However, it is also clear that these migrating beavers had difficulty finding unoccupied, suitable habitat. There is every indication that this pressure will continue.

According to Sieber (1997) about 400–500 beavers live in Austria along the rivers Danube and Morava. Zima (1989), using population genetics models, suggests that for the creation of a permanent population of viable animals (tests were done on mammals) a minimum of 500 individuals are necessary. The number of beavers in the floodplain forests of the Danube and Morava regions, as well as in appropriate habitats of the Zahorska lowland, is close to this number. It is important to note that smaller numbers of beavers penetrated through the Morava River to the Dije River as well. These populations are under the management of our Czech colleagues. We have directly observed and documented the origin and expansion of beaver populations in an area that covers parts of three countries. If the beavers living on the Hungarian side of the Danube River arm system are included then we have observed the origin of populations in four countries. While this region has a dense human population, and the land and rivers have been modified by human activities, beavers have still been able to form viable populations. When humans allow beavers to recolonize the landscape they have illustrated the ability to adapt to the new environment.

4.2. Danube Floodplain Forests

Beavers migrated down the Danube River and the Morava River and its tributaries, to settle in the floodplain forests on the Danube River arms. This part of the expanding beaver population is affected by several human-induced factors which are extremely significant. This region of the Danube was inhabited by beavers during the prehistoric and historic period. There were also many floods on the Danube during these times, however they did not appear to harm the beaver population. Because of this, it is necessary to examine some relevant circumstances related to the current water regime on the Danube, and its effect on the wildlife of the region. After its break through the Low Carpatian Mountains below the village of Devin, the Danube splits into numerous larger and smaller arms (originating and dissolving all the time). This area is bordered by half of the city of Bratislava, and by the mouth of the Vah River near Komarno. Matej Bell (1735) described the Danube in that day as splitting into four main arms (such as Moson and Kapltlne), which after its reconstruction, created the main river channel. The other main arms were the present-day Small Danube River (Maly Dunaj), and the present-day Cierna voda. Gravel lakes near Senec are also evidence that the Danube floodplain covered an even larger area in prehistoric times. It was important for beavers, as well as all other fauna of the floodplain forests, that any flood in this region caused only slow and gradual rising of the water level. This would allow all the animals time to find safe refuges when the water began rising.

This historic floodplain area has been reduced to a width of between 3–6 km after the regulation of the Danube, and closed with cross-barriers against floods. Obviously, in such conditions, water can rise 3–6 m every 24 to 36 hours during a flood, and the historic Danube has been changed into an enormous river of cold, muddy water tearing down everything in its path. These floods, which stopped as fast as they began, caused the inundation area to be changed into a dry desert during drier seasons.

The Gabcikovo Dam, completed in 1993, is designed to divert water from the main riverbed to a huge diversion canal, which carries the water to a turbine for the hydroelectric station. Construction of the dam eliminated the disastrous floods in the Danube River arms, however, the system of barriers near Cunovo, as well as the artificial filling of sections of the river arm-system, breaks the continuity of the Danube River. Aquatic and semiaquatic animals are now unable to use the river as a migration corridor, and can not reach the arm-system floodplain and colonize this area. The beaver density in the Rusovce arms in front of the Cunovo dam is increasing since beavers are prevented from continuing their movement down the Danube. It is possible that by traveling around the Low Carpatian Mountains these animals can settle in the arm-system. After the dam-system eliminated the destructive floods the arm-system region developed favorable hydrologic conditions for beavers. It is evident that the current dam-system requires further analysis and modification to function effectively as a wildlife corridor. One possible solution would be the construction of a bypass around Cunovo dam so beavers could gain access to the floodplain forests of the Danube River arms.

5. CONCLUSION

Forty-two beavers reintroduced in Austria were the founders of the current population of approximately 500 animals living in the Danube River and Morava River region. Beavers expanded their range inland to the Zahorska lowland by the Morava River and its left-side tributaries. This colonization confirmed that beavers are able to settle down and prosper in a human-modified landscape and in a highly developed country. We observed that beavers are selective in the way they colonize available habitat. Initially beavers occupy the most optimal sites, and then later migrants or dispersing animals will occupy suboptimal habitats. Once beavers are established in an area they begin to reshape the biotopography by constructing dams and lodges. We have also observed that large-scale water control developments on the Danube are insurmountable barriers for beavers. These hydrologic projects create a highly regulated river which distorts the natural water flow and flood pattern. This, in turn, influences the ability of beavers to maintain long-term residency in the region. The Gabcikovo Dam helped regulate floods, and also created optimal beaver habitat in the river arm system. However, the dam limits beaver movement along the Danube, and prevents beavers from entering the river arm-system. This environmental problem can be solved by a bypass around the Cunovo barrier.

Additional beavers, released in Poland, passed the ridge of the mountain Vychodne Beskydy. Several individuals now live on the Slovakian side, on the Southern banks of the mountains. We have also observed that a few beavers (2–4 individuals) crossed overland, through the ridge of the Low Carpatian Mountains near Brezova pod Bradlom, and then penetrated to the south of the Danube hollow.

REFERENCES

Bell, M., 1735, Noidia Hungariae novae historico goegraphica L

Budajova, J., 1995, European beaver (*Castor fiber* L., 1758) is part of Slovakia s fauna again (in Slovak), *Folia venatoria* **25:**175–181.

Chernel, I., 1886, Beavers in the Czallóköz (Zhitny island) (in Hungarian), *Vadtaszlap* **7:**257–258.

Kollar, H. P., Seiter, M., 1990, *Biber in den Donau-Auen östlich von Wien. Eine erfolgreiche Wiederansiedlung, Umwelt*, Verein für Ökologie und Umweltforschung, Wien.

Luknis, M., Plesnik, P., 1961, *Lowlands, valleys and mountains of Slovakia* (in Slovak), Osveta, Bratislava.

Pachinger, K. 1994, About the problems of living of the beaver in the system of Danube arms between Devin and Palkovicovo (in Slovak). Vyskum a ochrana cicavcov na Slovensku, p. 63–64.

Pachinger, K., 1995, *Occurrence and life perspectives of beavers in the Slovak section of the Danube River. Gabcikovo, part of the hydroelectric power project - environmental impact review*, Faculty of Natural Sciences, Comenius University, Bratislava, Slovakia, p. 31–39.

Pucek, Z., Raczynski, J., 1983, *Atlas of distribution of mammals in Poland* (in Polish), Pan, 188 pp. + 183 pp.

Sieber, J., 1997, Biber in Österreich: allzu erfolgreiche Heimkehrer, in: *Proceedings of the First European Beaver Symposium*, Bratislava, Slovakiam (unpubl.).

Somogyi, J., 1985, Beaver close to Bratislava (in Slovak). *Ziva* **33:**234.

Stollmann, A., Voskar, J., 1989, Return of the beaver into the landscape of Slovakia (in Slovak). *Folia Venatoria* **19:**245–256,

Tarnawski, A., 1987, Beavers in Beskydy Mountains (in Poland). *Przyroda Polska* **10:**15.

Zima, J., 1989, Methods of determining decrease in genetic variation in small populations (in Slovak). *Spravy* USEB **1989:**65–69.

9

BEAVERS IN AN URBAN LANDSCAPE

The Recent Activity of Beavers, *Castor fiber*, in the Greater Bratislava Area

Karol Pachinger[1] and Tomas Hulik[2]

[1]Institute of Ecology
[2]Department of Ecosozology and Physiotactics
Comenius University
Mlynska dolina B-2, 84215 Bratislava, Slovakia

ABSTRACT

Beavers immigrated to the Danube River near Bratislava in the 1980s. From 1984 to 1998 beavers have used 12 different locations for both long-term and short-term occupation in the Greater Bratislava area. We have developed a habitat classification system for use in an urban environment which includes human disturbance variables. In our system we consider the following habitat characteristics: 1) abundance of food, 2) water depth, 3) bank profile, 4) distance to a pedestrian walkway, 5) distance to a vehicular roadway, 6) degree of disturbance by humans, 7) duration of settlement, and 8) the number of times a location is resettled. Locations farther from Bratislava had higher habitat ratings, and were occupied for longer periods of time. Locations that are subjected to a high degree of human disturbance did not support long-term beaver settlements. While small-scale, local, human activity can influence beaver behavior, our observations indicate that large-scale human activities, such as the Gabcikovo Dam, have an even larger impact on the beaver population in the Greater Bratislava area.

1. INTRODUCTION

Pachinger and Hulik (this volume), describe the distribution and general size of the current beaver population in Slovakia. They report that beavers have been immigrating from reintroduction sites in Austria to the Morava River and into the Zahorska lowland in

Beaver Protection, Managment, and Utilization in Europe and North America, edited by Busher and Dzięciołowski.
Kluwer Academic / Plenum Publishers, New York, 1999.

Slovakia. Beavers have also spread along the Danube River below Vienna and into Bratislava, the capital city of Slovakia. We have observed that beavers have the ability to survive in both the outer, less developed, areas of Bratislava, as well as in the developed inner city. In these areas beavers are subjected to intense optical, acoustical, and physical disturbances which can limit their ability to persist in this region. In this paper we document the colonization of the Bratislava area by beavers and provide a descriptive analysis of the suitability of the habitat in the area. We also provide the first habitat classification system for beavers in an urban environment.

2. METHODS

Data were collected by direct observation. We recorded the length of time an area was occupied, the number of beavers in a settled area, the quality and quantity of the available food, the type of disturbance factors, and the overall quality of the environment in the inhabited location. With these data we provide an analysis of the habitat suitability in this urban environment.

2.1. Description of the Area

The Danube River enters outer Bratislava from the northwest on the Slovak-Austrian border below the village of Devin. After flowing through Bratislava the Danube exits the area near the Cunovo hydroelectric station. At Cunovo there is a barrier which diverts the Danube waters to a rafting canal, and to the hydroelectric station in Gabcikovo. Seven river arms (branches), that were reconstructed, surround the main river channel where the Danube passes through Bratislava. Three of these arms are connected directly to the main channel (Vodarenske, Zuzana, Home Rusovecke), and four are separated from the main channel by barriers, or by wide earth dikes (Jarovske, Kopacske, system of two lower Rusovce arms). We also examined the part of the Morava River which occurs before the confluence with the Danube River, and flows next to the former village of Devinska Novd Ves, which is presently one of Bratislava's boroughs. In this area we investigated the side arms of the Morava River, and the old gravel holes which are found here. The Rusovce River arms are partially backwaters of the original Danube River arms and are connected to the canals that were created by the construction of the Gabcikovo Dam.

2.2. Habitat Classification System

A very modest food base, and the presence of humans and nearby roads or railways are common aspects of most of the locations beavers occupy in Greater Bratislava. In order to analyze the environment in which beavers were observed we created an evaluation scale for classifying the habitat. This scale employs six basic environmental factors to determine the quality of each site beavers occupied. In order to avoid any complications caused by subjective evaluation methods, we designed a three point classification system (one point for the lowest quality, two points for average quality, and three points for optimal quality). Each of the six environmental factors was scored using this system. We also considered two additional factors: (1) the length of time beavers used an area, and (2) the number of different times (resettlement) an area was used, in our evaluation. With our system, considering only the first six factors, the very poorest habitat could have a low score of 6, and the very highest quality habitat could have a maximum score of 18. Adding the

Table 1. The eight factors and specific point criteria used in evaluating the quality of beaver habitat in Greater Bratislava

1.	Food Abundance—(herbaceous and woody vegetation)
	1 point—low abundance
	2 points—moderate abundance
	3 points—high abundance
2.	Depth of water
	1 point—50 cm to 1.5 m
	2 points— >3 m
	3 points—1.5 m to 3 m
3.	Bank profile
	1 point—a flat bank with a slope <40°
	2 points—a high bank with a slope from 40° to 60°
	3 points—a high bank with a slope from 60° to 80°
4.	Frequency of disturbance by direct human visits
	1 point—daily many times
	2 points—once every 2–3 days
	3 points—less than once per week
5.	Distance to a walkway (sidewalk)
	1 point—1–20 m
	2 points—20–80 m
	3 points— > 80 m
6.	Distance to a busy road
	1 point—10–40 m
	2 points—40–200 m
	3 points— > 200 m
7.	Length of occupation
	1 point—1–3 months
	2 points—3–18 months
	3 points— >19 months
8.	Number of times the site was resettled after the initial occupation
	1 point—one time
	2 points—two times
	3 points—three times

seventh factor into the evaluation (length of occupation), the lowest possible score is 7 and the highest is 21. When all eight factors are considered the range becomes 8–24. However, it is important to keep in mind that a site that is continuously occupied can only have a maximum score of 21 (factor 8 is not used since the site was never abandoned and resettled). Thus, with our system the most optimal sites could have scores of either 21 (continuously occupied) or 24 (repeatedly resettled). Table 1 outlines the classification system.

2.2.1. Aspects of the Environmental Factors. Our evaluation of the factors outlined in Table 1 included the following considerations.

1. In our classification of food abundance we attempted to evaluate the amount and quality of the vegetation available for use by beavers. Aquatic plants and plants growing in the riparian area must be abundant enough to provide for the nutritional needs of the animals during the vegetative period (spring, summer, autumn). Beavers are generalist herbivores (see the papers by Basey, and Schulte and Müller-Schwarze in this volume), and often supplement their normal diet with agricultural products (for example, corn and beets) when these are

available. During the winter beavers use woody vegetation (willows and poplars). We evaluated each location and estimated the availability of both summer and winter foods. We recognized that a low abundance of herbaceous vegetation may be balanced by the use of woody vegetation (including the leaves of trees and shrubs) during summer. When beavers use mainly trees and shrubs during the summer they open up the floodplain forest by cutting down trees, and herbaceous vegetation is more abundant in the following years. This can affect the long-term occupation of a site. However, beavers do need an adequate supply of woody vegetation during the winter or survival is doubtful. The abundance and quality of food is a limiting factor to beaver colonization.
2. Water depth is also a critical factor for beaver colonization. When there is insufficient water, or an unstable water level (if other environmental factors are optimal), beavers start changing their environment by building dams. Our observations indicate that beavers did not settle in areas with low or fluctuating water levels. The ideal water depth is between 1.5 m to 3 m. This depth provides beavers with enough water to escape from danger, and also allows light to reach the river bottom enabling the growth of aquatic vegetation. Water that is too shallow or too deep is not considered optimal in our system.
3. We also consider the bank profile a limiting factor influencing colonization of an area. A high bank with a relatively steep slope appears to be important for beavers for the construction of bank burrows. Our system primarily differentiates between flat bank profiles and high profiles. In the area of Greater Bratislava beavers usually settled only in locations where they could easily construct bank burrows.
4. Human Disturbance is also critical in an urbanized area like Greater Bratislava. If human activity is limited to simply examining the cutting activity around the beaver lodge, beavers adapt by being active only at night. However, if people attempt to force the beavers out of the lodge and try to capture them this can cause the animals to leave the location. In our classification system 1 point is awarded if human disturbance has forced the animals to limit their activity to the night. This is a potential problem since during the summer, nights are of short duration and beavers may not obtain enough food. We only consider human disturbance that is indirect (the examination of the area). Direct attacks are usually not observed and we are aware of them only after the animals disappear. Our system considers the frequency of people in the beaver area.
5. The distance to both walkways and automobile roads can influence beaver activity. Proximity to a walkway increases the likelihood of human disturbance. Automobile roads present other problems. While beavers seem to acclimate to the noise coming from a road, there is the increased probability of beavers being killed by automobiles, and also the effect of environmental pollution from automobile exhaust. However, we have not recorded any beaver mortality caused by a motor vehicle during our 14-year study. Proximity to a road also increases the potential for human disturbance. In our system optimal beaver habitat is considered to be farthest from both roads and walkways.
6. In our system we also examine the period of time beavers occupy an area, and the resettlement of areas that are abandoned. Optimal locations are ones that are occupied for longer periods. Resettlement of a location also may provided insight into what is optimal habitat. For example, a site that is continually resettled after beavers are expelled, is considered more optimal than one that is not.

3. RESULTS

3.1. Location and Duration of Beaver Activity

Since beavers were first observed in the Greater Bratislava area in 1984 we have been documenting their activity. Beavers have used 12 different locations for both long-term and short-term occupation (Figure 1). The length of time beavers occupied a specific location ranged from 3 weeks to 7 years. We estimate that during the past 14 years 25 beavers have lived temporarily, and 13 beavers have lived permanently in the Greater Bratislava area. However, these numbers are only estimates since we could not always determine how many beavers were present at each location. It is possible the actual numbers could be larger (4–6 additional animals). The dates the areas were occupied, and the duration of each occupation are presented below.

Location 1: Arm Zuzana. This area was used three times: (1) February 1984 to April 1985, (2) the winter of 1993/1994, and (3) from spring 1997 to autumn 1997.

Location 2: Vodarensky Island I. This site has been occupied since 1991, and is still occupied at the present time.

Location 3: Kopac River Arm. Beavers have been observed in this area twice in the past year, and it is occupied at the present time.

Location 4: The Morava River near Sandberg hill (Devinska Kobyla National Nature Reserve).

Location 5: Rusovce Lake. This site was occupied during the winter of 1995/96.

Location 6: Rusovce River arm. Beavers have used this site continuously since the winter of 1996/97. It is currently occupied.

Location 7: In Petrzalka near a mouth of water treatment facility. Beavers used this area for a short-term stay during the summer of 1996.

Location 8: Biological Research Station Jursky Sur. This area has been used twice: (1) From October 1996 to July 1997, and (2) from November 1997 to April 1998.

Location 9: Chorvatske River arm, inner Petrzalka. This is a former river arm that has been reconstructed into a drainage canal. It was used for a short-term stay during March/April 1997.

Location 10: Devinska Nova Ves—Marchegg bridges. Beavers have occupied this site since autumn 1996. It is still occupied at the present time.

Location 11: Unnamed small island in the Vodarenske River arm. This area has been occupied since the spring of 1997.

Location 12: Karlova Ves bay—near the shipyard. This area was occupied for a short time during January 1997.

3.2. Habitat Classification of Occupied Sites

The total habitat quality scores for the 12 beaver sites ranged from 7.0 to 21.0 with an average score of 15.5 (Table 2). The highest quality sites were on the outskirts of the city on the Danube islands (refer to Figure 1). The site with the lowest score (location 9; 7 points) was situated close to the city. Three of four sites closest to Bratislava (Locations 7, 9, 12) were only used on a temporary basis (short-term) and had an average score of 9.3. The average score for the five sites that are currently occupied (locations 2, 3, 6, 10, 11) is

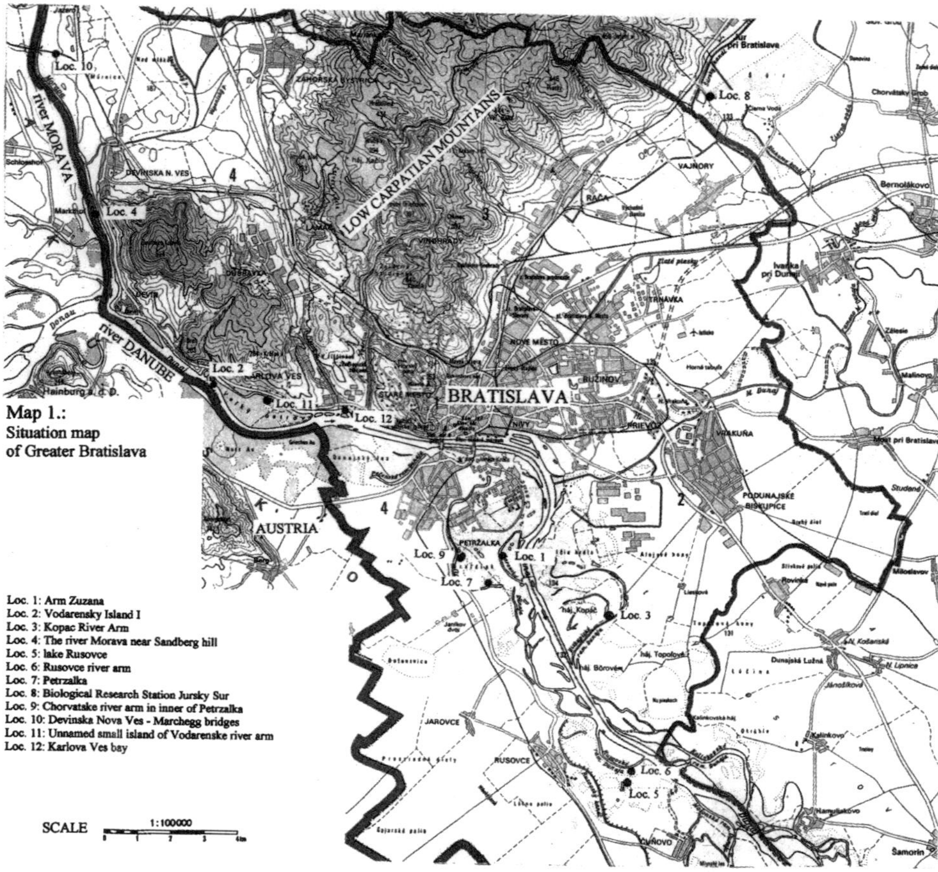

Figure 1. Map of the Greater Bratislava area showing the locations beavers have occupied in the past 14 years.

19.5. The average score for the four sites (locations 5, 7, 9, 12) that were only occupied once, and for a short time, is 11.25. While we did not statistically analyze the differences between these two groups of locations it is apparent that they are very different in quality. At a number of locations beavers had to abandon the territories because of unfavorable external conditions. Three of these abandoned sites were recolonized at a later time, and these three sites had an average habitat score of 17.7.

4. DISCUSSION

Why do migrating beavers settle in the Greater Bratislava area? Why do beavers choose locations that are noisy and near busy roadways when they could have found more peaceful sites along other sections of the Danube? These are difficult questions to answer. If we examine locations that are colonized, abandoned, and then recolonized it may be useful. At one site beavers were initially forced to leave due to a flood. The next time they had to leave because of disturbance by people, and possibly because of poachers. Presently, this location is inhabited for the third time regardless of continual disturbance by humans. We did not find out what specific factor makes this location attractive for bea-

Table 2. Habitat quality at locations occupied by beavers in the Greater Bratislava area. Habitat characteristics are described in the text and in Table 1

Location	Food	Water depth	Bank profile	Roadway	Walkway	Disturbance	Duration of stay	Resettlement	Mean
1	2	3	3	1	1	1	2	2	15
2	3	3	3	3	2	3	3	1	21
3	3	3	3	3	3	2	2	1	20
4	2	3	2	1	1	2	1	—	12
5	1	3	3	3	1	2	1	—	14
6	3	3	3	3	1	2	3	3	21
7	3	1	2	3	1	1	1	—	12
8	2	3	3	3	1	2	3	1	18
9	1	1	1	1	1	1	1	—	7
10	3	3	3	3	1	2	2	1	18
11	3	1	2	1	3	3	3	—	16
12	1	3	2	3	1	1	1	—	12
Mean	2.3	2.5	2.5	2.3	1.4	1.8	1.9	0.8	15.5

vers. We can only speculate that the bank profile and the bank burrow the beavers use acts as a strong attraction for the migrating animals. We experienced a very similar phenomenon in the area of the Kopac River arm (location 3 in Figure 1). In this area the beavers always inhabit the same bank burrow. Here we have observed that beavers examine the entire area and then show great selectivity in choosing their permanent burrow. We have also observed an aborted stay in a gravel hole and in dead branches of the Morava River near Devinska Nova Ves (near the Biological Research Station in Sur). This behavior suggested a loyalty to a place which had previously been abandoned by beavers due to low water levels. Site selection by dispersing beavers appears to be multi-faceted and demands more study.

Our observations suggest that in most locations in the Greater Bratislava area beavers intended to settle permanently, not just for a short time. We did identify temporary settlements in some locations, but at most locations that were used temporarily, beavers were forced to leave due to lack of food, poaching, or low water level. At these sites beavers returned as soon as the situation allowed them to do so. While beaver habitat quality is difficult to accurately determine, an adequate food supply coupled with a relatively low degree of human disturbance are critical factors. For example, the longest stay we have observed is at location 2 where beavers have been living since 1991. In the beginning of their occupation the beavers survived large floods which inundated almost the entire island. The beavers somehow survived the flood and, since this site has abundant food throughout the year and is high quality habitat, it has been continuously occupied since that time. Human disturbance is low at this site. A second location which illustrates this same pattern is the small island in the branch of the Vodarenske River (location 11), first occupied in 1997, and still inhabited at the present time. At this site the beavers are not disturbed by people, seem to have adapted to the noise coming from a nearby road, and have an abundant food supply. Beavers did not persist in areas with a high amount of human disturbance. The sites with the most disturbance (locations 1, 7, 9, 12) did not support long-term settlement. Location 1, which has more food, and more favorable water depth and bank profile than the other three locations, was used more often by beavers. However, beaver-human interaction appears to be a critical factor that influences long-term settlement of locations in the Greater Bratislava area.

In addition to local, relatively small-scale disturbances by humans, beavers are affected by large-scale human activities. In 1993, completion of the Gabcikovo Dam on the Danube dramatically altered the situation for beavers living downstream from Bratislava. Dispersal downstream along the Danube was completely prevented by the dam, and by 1997 beavers overcrowded the area of the Rusovce River arm system. Also, the entrance into the rafting canal has slippery banks which prevent beavers from leaving the water. If beavers enter the canal, which they do attempt to stay away from, it means almost certain death. We were informed that during the period of high water in the spring and summer of 1997 five beavers were found dead near the hydroelectric generator. Thus, the dam and its associated water channels pose a direct threat to beavers in this area of Greater Bratislava.

7. CONCLUSION

We have developed a beaver habitat classification system for use in an urban area. While our classification system is simple, it is also very easy to use, and based on these results is able to clearly differentiate between high quality and low quality habitats in an urban area. Based on the results of 14 years of study we consider most habitats in the Greater Bratislava area to be either average quality or poor quality, which will limit the ability of beavers to permanently colonize the area. Our classification system, with slight modifications for local conditions, can be used in other urban areas. It can also be used to predict which currently unoccupied areas may be suitable for beavers in the future, and how increased human development will impact potential beaver habitat.

10

A BIOCHEMICAL-GENETIC DISCRIMINATION METHOD FOR THE TWO BEAVER SPECIES, *CASTOR FIBER* AND *CASTOR CANADENSIS*, AS A TOOL FOR CONSERVATION

Johanna Sieber,[1] Franz Suchentrunk,[2] and Gunther B Hartl[3]

[1]Konrad Lorenz Institute for Comparative Ethology
Austrian Academy of Sciences
Savoyenstrasse Ia, A-1160 Vienna, Austria
[2]Research Institute of Wildlife Ecology
Vienna Veterinary University
Savoyenstrasse 1, A-1160 Vienna, Austria
[3]Institut fur Haustierkunde
Christian Albrechts Universität zu Kiel
Olshausenstrasse 40-60, D-24118 Kiel, Germany

ABSTRACT

European beavers, *Castor fiber*, were recently reintroduced to Austria and Bavaria and the population has been increasing rapidly. In 1996 we began livetrapping animals for reintroduction programs in neighboring countries. There are currently North American beavers, *C. canadensis*, in Bavaria, and a small number were inadvertently released in Austria 20 years ago. It is critical to be able to discriminate between North American beavers and European beavers that are trapped as part of any reintroduction program in Europe. For our reintroduction work we need to definitatively determine the species of each trapped beaver so no new North American beavers are released. It is also critical to be able to discrimiate between the two species so the distribution of North American beavers in our region may be determined. Discrimination between the two species by morphometric methods is difficult and not definitive using live animals. Preparing chromosomes for karyotypic studies takes time and only works properly with fresh blood. Therefore, we developed a method that uses tissues from either live or dead animals as well as fresh or deep-frozen blood. Earlier, using multi-locus allozyme electrophoresis to

Beaver Protection, Managment, and Utilization in Europe and North America, edited by Busher and Dzięciołowski.
Kluwer Academic / Plenum Publishers, New York, 1999.

compare two North American beavers and two European beavers, we only observed variation at the Esterase-D (E.C.3.1.1.1) locus out of the 14 loci examined. To further test our new method of discriminating between the two species we used either animals whose species was already known, or confirmed our results by preparing additional karyotypes. We screened blood and kidney samples from 37 individuals of *C. fiber* and from 9 individuals of *C. canadensis* for variation of the Es-D locus by horizontal starch gel electrophoresis. *All C. fiber* exhibited a fast migrating allele and all *C. canadensis* a slow migrating allele at this locus. Hence, the Es-D locus can be interpreted as a differential-diagnostic for the two species.

1. INTRODUCTION

1.1. Background

After complete extirpation in the middle of the last century (von Mojsvar, 1897) beavers were recently reintroduced to Austria and Bavaria. These reintroduced populations have been increasing rapidly. (Kollar and Seiter, 1990; Schwab, Dietzen, and Von Lossow, 1994; Sieber, 1995, 1998). The introduced animals originated from different European beaver subspecies (*Castor fiber fiber*, *C. f. osteuropaeus*, *C. f. albicus*, *C. f. vistulanus*). Unfortunately, some North American beavers, *Castor canadensis*, which are not endemic to Europe, were also released (Koenig and Krebs, 1979). Some of these North American beavers were released legally (although they were not identified as *C. canadensis* at the time of the introduction), while others were released illegally by private persons, or when animals escaped from animal parks. Evidence from Finland suggests that North American beavers reproduce more successfully, and their populations increase more rapidly than European beavers when the species are sympatric (Ermala, Helminen, and Lahti, 1989). Therefore, there is a risk that North American beavers would partly or completely displace European beaver populations after some years. It is vital, in an effort to aid the reintroduction of European beavers, to locate *C. canadensis* individuals in central European beaver populations and to remove them. In an attempt to find a fast working, reliable, and inexpensive discrimination method that is applicable to both dead or live animals, we screened allozyme loci from both beaver species for diagnostic allelic variation.

1.2. Previous Species Discrimination Methods

1.2.1. Morphometric Method. Even though. *C. canadensis* and *C. fiber* are superficially similar, and thus are hard to discriminate in the field, the two species do show significant differences in their anatomy (Lavrov, 1974). The most distinctive differences are found in the skull. North American beavers have shorter os nasale with blunt-edged ends, the foramen occipitale are more oval shaped, and the foramen nasale are trapezoidal. European beavers have longer os nasale with pointed ends, the foramen occipitale are circle shaped, and the foramen nasale are triangular. The major disadvantage of this method is that the differences are only easy to judge when measuring skeletons and not live animals.

1.2.2. Genetic Method. The two species are easy to distinguish when comparing chromosomes since C. *canadensis* has a diploid number of 40, while *C. fiber* has a diploid number of 48 (Hsu and Benirschke, 1968; Lavrov and Orlov, 1973). However, this method only works properly when fresh blood can be used. Additionally, this method is time consuming, expensive, and requires a well equipped laboratory.

1.2.3. Anal Gland Secretion. Rosell, Sun, and Bergan (1997) presented a method using the color of the anal gland secretion for discrimination between North American and European beavers. *Castor canadensis* males and *C. fiber* females have different color secretions, but *C. canadensis* females and *C. fiber* males have similar color secretions. Therefore, this method can only be used if the sex of an animal is known and fresh secretions can be sampled.

2. METHODS

2.1. The Protocol

We have developed a method that can use tissue from either live or dead animals, as well as fresh or deep-frozen blood. Additionally, our method is both rapid and easy to perform which permits short handling and maintenance times. This is critical when beavers are being used in reintroduction or translocation programs. In an earlier study comparing two individuals of each species we examined 14 loci using multilocus allozyme electrophoresis and observed only variation at the Es-D locus. To further test our new method we used either animals of known origin or confirmed our results by preparing additional chromosome patterns. In the case of dead animals we examined skull characteristics such as the position and shape of sutures of the nasal bones, and the shape of the foramen magnum to confirm the species diagnosis (Niethammer and Krapp, 1978). The allelic variation at the Es-D locus (E.C.3.1.1.1) was analyzed by conventional horizontal starch gel electrophoresis (Rothe, 1994). We used potato starch, hydrolyzed for electrophoresis (Sigma), for producing 12.5% gels, and a Tris/Citrat buffer (pH 8.0). Gels were run for 5.5 h at 130 V and 95 mA. In the staining procedure we used 4-Methylumbelliferyl acetate (Richardson, Baverstock, and Adams, 1986), and gels were viewed under UV- light.

2.2. First Application

In 1992 a few small beaver family groups were found in southeastern Austria. These groups had no connection to the well established population in northeastern Austria and we assumed that they were descendants of North American beavers which had recently escaped from a zoo. As this had to be proved before the animals were trapped and removed, we used our new method to screen blood from two animals. The results of the test indicated that they were *C. canadensis*. Subsequently we live-trapped the animals and gave them to different zoological parks.

2.3. Further Application

Last year we started to livetrap animals for reintroduction programs in neighboring countries. As a small number of North American beavers, together with a much larger number of European beavers, were released some 20 years ago, we need to determine the exact species of each trapped beaver to avoid further spread of North American beavers. As *Castor fiber* and *C. canadensis* do not interbreed, it was only necessary to determine

Figure 1. Map illustrating the locations of the populations that were sampled.

the species of one member of each family we livetrapped. We then assumed that all the other family members were the same species as the animal we tested. Additionally, we are interested in determining the species of any carcass that is found, since this will allow us to determine the distribution of any North American beavers that remain in the wild.

3. RESULTS

3.1. Allelic Differences

Electropherograms of a total of 37 European and nine North American beavers were obtained from locations in Austria and Germany (Figure 1). All animals screened of both species were from different specific sites, and hence there is a low likelihood that any individuals were related. All individuals showed a single band pattern, respectively. This corresponds to a homozygote genotype in each individual for this dimeric enzyme (Richardson *et al.*, 1986). All European beavers exhibited a slow migrating allele and all

North American beavers a fast migrating allele. This result supports our earlier view of differential diagnostic alleles at the Es-D locus for this pair of species.

3.2. Conservation Application

Since 1996 we have tried to determine the species of every Austrian beaver that is livetrapped or found dead. Additionally, we had the chance to examine live beavers and carcasses from Bavaria and individuals of the subspecies *C. fiber albicus*. We found North American beavers at only two locations: the escaped beavers mentioned above (see section 2.2), and another two individuals which were killed by people at a second site. We suspect that these two beavers had originally been illegally released in Bavaria, and then dispersed into Austria. Thus, we did not find any evidence of an increase in, or spread of, a North American beaver population in Central Europe. However, there is the possibility that North American beavers may have survived in undetected locations, and we continue to use our technique to examine any beaver (dead or alive) that is brought to us.

REFERENCES

Ermala, A., Helminen, M., and Lahti, S., 1989, Some aspects of the occurence, abundance and future of the Finnish beaver population, *Suomen Riista* **35:**108–118.

Hsu, T. C., and Benirschke, K., 1968, *An Atlas of Mammalian Chromosomes*, Vol. 2, Springer Verlag, New York, Heidleberg, Berlin.

Koenig, 0., and Krebs, U., 1979, *Bedeutung und Methodik der Ansiedlung von Bibern*, Forschungsgemeinschaft Wilhelminenberg Wien.

Kollar, H. P., and Seiter, M., 1990, *Biber in den Donau-Auen östlich von Wien: Eine erfolgreiche Wiederansiedlung, Umwelt* **14**. Verein für Ökologie und Umweltforschung, Wien.

Lavrov, L. S., 1974, Morphological types of Palaearctic beavers of Castor genus and some aspects of their taxonomy, *Transactions* **1:**344, lst Intern. Theriol. Congress, Moscow,1974.

Lavrov, L. S., and Orlov, V. N., 1973, Karyotity I taksonomija sovremennich bobrov (*Castor*, Castoridae, Mammalia), *Zool. Zhur*. **52:**734–742.

Niethammer, J. and Krapp, F., (eds.), 1978, *Handbuch der Säugetiere Europas*. Volume 1/1, Akademische Verlagsgesellschaft, Wiesbaden.

Pachinger, K., 1994, Spontane Verbreitung des Bibers im Flußtal der March und der Zahorska Tiefebene. *Ekologia Bratislava* **1994:**Suppl. 1.

Rahm, U., and Baettig, M.,1996, Der Biber in der Schweiz. Bestand, Gefährdung, Schutz, *Schriftenreihe Umwelt* **249**. Hrsg. Bundesamt für Umwelt, Wald und Landschaft.

Richardson, B. J., Baverstock, P. R. and Adams, M., 1986, *Allozyme Electrophoresis. A Handbook for Animal Systematics and Population Studies*, Academic Press Inc., San Diego.

Rosell, F., Sun, L. Bergan, F., 1997, Use the colour of the anal gland secretion to differentiate between the two beaver species, in: *Proceedings of the First European Beaver Symposium* , Bratislava, Slovakia, (unpubl.).

Rothe, G. M., 1994, *Electrophoresis of Enzymes. Laboratory Methods*, Springer LAB MANUAL, Springer Verlag, New York, Heidelberg, Berlin.

Schwab, G., Dietzen, W., and Von Lossow, G., 1994, Biber in Bayeern: Entwicklung eines Gesamtkonzeptes zum Schutz des Bibers, *Schriftenreihe Bayerische Landesamt für Umweltschutz* **128:**9–31.

Sieber, J., 1995, Sie schwimmen wieder ! Biber (*Castor fiber*) in Österreich, *Stapfia* **37:**217–224.

Sieber, J., 1998, 20 Jahre nach der Wiederansiedlung. Biber-Management unvermeidlich? *Naturschutz und Landschaftspflege in Brandenburg* **1:**79–80.

von Mojsvar, A. M., 1897, *Das Thierleben der Österreichisch-Ungarischen Tiefebene, Biologische und thiergeographische Skizzen und Bilder*, Alfred Hölder, Wien.

11

ECOLOGICAL RESTORATION BY HARNESSING THE WORK OF BEAVERS

Yurii A. Gorshkov,[1] Andrea L. Easter-Pilcher,[2,*] Brian K. Pilcher,[2] and Dmitry Gorshkov[1]

[1]Volga-Kama National Preserve, Sadovaya
P.O. Raifa, Zelenodolsk District, Tatarstan, 422523 Russian Federation
[2]Western Montana College of the University of Montana
710 S. Atlantic, Dillon, Montana 59725

ABSTRACT

The Volga-Kama National Preserve in Tatarstan, Russia, has been undergoing subtle but steady changes in recent decades due to hydrological impacts. Sediment from upstream agricultural lands is causing lakes and bogs to fill. The landscape is also drying, causing losses of boreal species (biodiversity). The drying may be due to lower off-peak stream flows from forest conversion to agriculture, which could cause a drop in the groundwater level. We have begun attacking the problem with solutions on two fronts and are collecting baseline data to monitor the success of our progress. One attack front involves the re-establishment of beavers. Their dams will catch sediment, lengthening the life of lakes and bogs. The dams will raise the level of surface water which, in turn, will raise the local groundwater level and provide a mesic environment for declining populations of boreal species. A second and longer-term front on which to solve the erosion problem is to find the sources of sediment and implement less-erosive agricultural practices. Baseline information will be obtained from permanent vegetation plots, remote sensing images, and the preserve's records of lake measurements.

---

* Principal investigator.

Beaver Protection, Managment, and Utilization in Europe and North America, edited by Busher and Dzięciołowski.
Kluwer Academic / Plenum Publishers, New York, 1999.

1. INTRODUCTION

1.1. The Landscape and Its Problems

The Raifa part (3,864ha) of the Volga-Kama National Preserve (VKNP) has been undergoing subtle but steady changes in recent decades that can be likened to the Chinese "death by a thousand cuts". The preserve is located 800 km east of Moscow and 600 km west of the Ural Mountains near the city of Kazan, Republic of Tatarstan, Russian Federation. Much of the VKNP forest is 300 years old and the preserve would be in good shape except for the influence of agricultural lands upstream. Thousands of hectares of forest north and southeast of the preserve were cleared in the last two centuries and converted to cropland. The two rivers that flow through the preserve (Sumka and Ser-Bulak) receive runoff from upstream agriculture, and the changes brought by these rivers threaten the entire hydrologic regime of this treasured relict.

These small rivers are effecting annually subtle, but cumulatively large, changes in two ways— through their sediment delivery, and lower off-peak stream flows. Their heavy loads of silt are filling in lakes and bogs. Due to the Sumka River, the deepest part of Raifa Lake filled in at the rate of 4 cm/yr between 1970 and 1990, but the worst sediment accumulation is at the inlet. The lake lost 3.6 ha, or 10% of its area. The combined effect has been a loss of 18.9% of lake volume. The impact would be worse if it were not for Lake Belloya, a lake upstream from Raifa and off of the preserve. In the process of intercepting sediment making its way toward Raifa, this lake has lost 77.5% of its volume over the same 20-year interval. Unfortunately, a tributary to the Sumka (Sapa River) enters the river below Lake Belloya, allowing an impact on Raifa. The Ser-Bulak River and its tributaries have carried sediment into lakes and bogs, too. Lake Lenyovo lost 14.9% of its volume in 20 years and Long Lake (and bog), though not measured, has new, significant deposits at its inlet each year.

Scientists of the preserve and Kazan State University (KSU) suspect that the preserve is being influenced by lower off-peak stream flows in these rivers. Loss of upstream forest by clearing for agriculture would cause off-peak stream flows to be less (Nikoli Torsuev, hydrology professor, KSU). This, in turn, would lower the adjacent water table. We have observed a drying trend on the preserve that has resulted in a loss of some boreal species and we fear the loss of more boreal species through this xerification. Data that will document our observations will be collected over the next few years. Some plants and animals have already disappeared or are likely to disappear from this part of the preserve in the near future— the northern redback vole (*Clethrionomys rutilus*), Russian flying squirrel (*Pteromys volans*), hazel grouse (*Tetrastes bonasia*), white fir (*Abies alba*), Finland spruce (*Picea finlandia*), an extremely rare Lycopodiaceae (*Huperzia selago,* formerly *Lycopodium selago*), a Gramineae (*Schizachne callosa),* a Pyrolaceae (*Moneses uniflora*), four orchids—*Malaxis monophyllos, Corallorhiza trifida*, *Gymnadenia conopsea,* and the extremely rare Calypso orchid (*Calypso bulbosa*). We are losing boreal habitats and biodiversity in general.

1.2. General Plan

We have begun to attack these threats and to monitor our effectiveness through several means. The first step was to establish our collaborative relationships to share expertise and resources. Yurii Gorshkov, Director of the Volga-Kama National Preserve, and Andrea Easter-Pilcher and Brian Pilcher from Western Montana College of the University of Montana (WMC), Dillon, Montana, U.S.A., are focusing on the ecological aspects of the problems. Martin Spetich of Purdue University, West Lafayette, Indiana, contributed to the formulation of methodology in remote sensing and forest sampling. We have begun working on the agricultural erosion sources and groundwater measurements with Dr. Nikoli Torsuev of Kazan State University, Kazan, Tatarstan.

The second step was to reintroduce beavers (*Castor fiber*) in an attempt to reverse or slow these damaging changes in the short-term. Beavers will slow lake and bog sedimentation and reverse the drying trend through the capacity of beaver ponds to capture sediment, raise water levels, and increase the distribution of water. The last of the beavers in the region were killed in 1802. Beginning in 1947, beavers were reintroduced to many parts of Tatarstan and the current population numbers approximately 1,000. However, only 16% of the republic is forested, and the island of forest in which the preserve is located would most likely not be recolonized soon, if at all, without human intervention.

We will monitor the effectiveness of beavers in attaining our ecological restoration goals. We are gathering baseline data through measurements of lake dimensions, permanent vegetation plots following (with slight modifications) plot methodology of Easter-Pilcher (1987), and classification of remote sensing images. We will be able to detect direct and indirect changes in siltation rates, changes in quantity of vegetation by taxon, and changes in water distribution.

We view the use of beaver activity as only a short-term solution to the larger sedimentation problem since beaver ponds will only hold a finite quantity of sediment. Our long-range objective is to work directly on sediment sources. After we locate sediment sources, we will work with the farmers to improve agricultural practices that will minimize erosion in those areas.

1.3. Beavers—an Attractive Aid

Reintroduction of native species into portions of their historic range has been recognized as a potentially powerful conservation tool (Hartman, 1994). Beavers have tremendous potential to modify their surroundings and so they provide an excellent example of how long-term ecosystem and landscape-level changes are brought about by herbivory, habitat alteration and population dynamics (Easter-Pilcher, 1987, 1990; Johnston and Naiman, 1990; Naiman and Melillo, 1984; Naiman, Melillo, and Hobbie, 1986; Naiman, Johnston, and Kelley, 1988). Forest ecosystem changes resulting from reintroduced beavers (25 pairs in 1946) in Tierra del Fuego, Argentina included a significant increase in wetland areas and a significant increase in the watertable level (Lizarralde, 1993).

Historically, beavers had significant effects on small to moderate sized streams throughout the north-temperate zone (Naiman *et al.*, 1988). Prior to extirpation, in many areas throughout this region, beaver ponds along stream ecosystems led to the supply and retention of coarse particulate matter (Naiman *et al.*, 1986), invertebrate production (McDowell and Naiman, 1986), and boundary complexity (Johnston and Naiman, 1987, cited in Schlosser 1995).

Beavers alter riparian ecosystems by felling trees, constructing dams, creating water impoundments, utilizing woody species for food and transporting woody materials from one location to another, thus altering woody and herbaceous species composition and causing horizontal and vertical stratification of woody biomass (Easter-Pilcher, 1987; Johnston and Naiman, 1987; Lizarralde, 1993). As a keystone species, beavers modify stream morphology and hydrology as well as biogeochemical pathways and community productivity (Naiman *et al.*, 1986). Wetlands are created and maintained, sediment and organic matter are retained in stream channels, nutrient cycling and decomposition dynamics are modified, riparian zone structure and dynamics are modified, and the character of water and materials transported downstream is influenced. Ultimately, the composition and diversity of the entire biotic community is affected (Collins, 1976; Naiman and Melillo, 1984; Naiman *et al.*, 1986; Sielfeld and Veneges, 1981, cited in Lizarralde 1993).

Diversity and composition of biotic communities are influenced by several significant factors (Hair, Hepp, Luckett, Reese, and Woodward, 1979). The availability of microhabitats increases as the structural heterogeneity of a habitat increases (Willson, 1974). This generally corresponds to an increased number of species in a given area (Emlen, 1973; Gauthreaux, 1978; Karr and Roth, 1971; Rogers, 1967). Beaver impoundments exhibit high structural heterogeneity and an associated high level of species diversity and as such are recognized as critical wetland habitat. Hair *et al.* (1979) documented significantly higher numbers of bird species in beaver impoundments than in adjacent hardwood and pine areas. Schlosser (1995) has suggested that beaver ponds may act as reproductive "sources" (sensu Pulliam 1988).while adjoining streams may act as reproductive "sinks" for fish. Johnston and Naiman (1987) hypothesize that "...preferential central-place foraging by beavers scattered throughout the landscape could establish a patchwork of shrubs and conifers in an otherwise deciduous forest matrix, increasing forest diversity."

2. BEAVER REINTRODUCTION

Our first step was to reintroduce beavers to the preserve. To date, founding animals have been trapped in beaver ponds of the Viatka River Plain of eastern Tatarstan in the summers of 1996 and 1997. More trapping is planned for late spring of 1999 and three of those new beavers will be implanted with abdominal transmitters. Beavers were weighed, their sex determined (as best we could), measured and ear-tagged. In 1996, one beaver was released on Raifa Lake, one on Lake Ilantovo, and a family group on Lake Lenyovo. The Lenyovo family group was put into an artificial bank den for a "soft" release. The den was constructed by digging a large pit connected to the lake by a trench. Logs were used to cover the pit and trench, and they in turn were covered with dirt. In 1997, two of the adult beavers from a second transplant family group were implanted with abdominal transmitters following methodology of Melquist and Hornocker (1979), and Davis, Von Recum, Smith, and Guynn (1984), and the family was put into the artificial bank den on Lake Lenyovo. To date, 13 beavers have been released on the preserve (6 adults, 1 subadult, 6 kits). There were 2 single adult females and 2 family groups. There have been 2 known post-release mortalities, one of which was from pneumonia, the other from unknown causes.

3. BEAVER ECOLOGY

3.1. Initial Dispersal

At least 2 to 3 beavers have remained on the preserve, while up to 8 (depending on unknown mortality) have moved upstream on both rivers 1–9 km outside the preserve boundary and up to 16km from the nearest release site. There are now 4 known occupied sites in the general area of the preserve. The 2 areas on the preserve are Ser-Bulak River near Lake Lenyovo and Lake Gniloye. The 2 areas off the preserve are the upper Sumka

River and eastern buffer zone. No single dispersal pattern characterized lone animals or families. One single female moved only 1 km while another moved 16 km. One family group moved 1 km while another family moved 4 km from the release site. The male split off from this second family group, but settled with its family the following spring.

3.2. Habitat Selection

Vegetation and other site characteristics will be measured in permanent, systematically located plots and transects on lakes and streams (methods adapted from Easter-Pilcher, 1987, with some additions). These will serve several purposes, including the characterization of preferred habitat. Additional plots and transects will be located if necessary to characterize colony sites as they are established. Baseline vegetation and habitat data were collected on 51 transects in July and August of 1997 and 1998. Roughly 15 more are planned for next year. Small plots will sample herbaceous species richness and relative abundance, medium-sized plots will sample woody stems cut by beavers or available to beavers by distance from water, and larger plots will sample forest overstory.

None of the initial habitat selection data has been compiled since we are still collecting data, but it is apparent that beavers have thus far settled in bog areas with floating vegetation mats and abandoned, flooded peat mines.

3.3. Utilization of Woody Vegetation

We began tracking forage utilization through Dmitry Gorshkov's study. By species, he counted woody stems cut and approximated the volume of wood utilized during August-November 1996. He found that 95% of the number of cut stems were European gray willow (*Salix cinerea*). As measured by volume, aspen (*Populus tremula*) was utilized the most (46%), followed by European gray willow (23%) and European white alder (*Alnus incana*). Birch (*Betula* spp.) stems were cut the most in the following spring and summer. Aspen may have been exhausted by this time, but explaining the switch to birch by the absence of aspen isn't entirely satisfactory, because the other species above were still available and were used in the fall to a much greater extent. However, it probably explains why aspen use was zero. The aspen was used to build caches, while the birch was used to build dams and possibly for some degree of consumption.

Information from Dmitry Gorshkov's study was utilized to calculate carrying capacity of the Ser-Bulak River Plain on the preserve. He found that 6 beavers utilized 3.9 m^3 of trees in the fall. Beavers consume 60% of their annual utilization in the fall (Lavrov 1934), so application of these consumption rates yields an estimate of 1.083 m^3/beaver/year. Annual growth is currently 2 m^3/ha on 14 ha of forested plain. These growth rates would support 26 beavers indefinitely on this plain. Although these tree growth rates are destined to change, it is safe to say that the activities of many beavers could be supported along this one river.

So far, the released beavers have tended to cache very little, leading to speculation about their foraging strategies. They have been active to some degree through the winter in cutting woody stems on the bank, and the floating vegetation mats and peat bogs provide access to roots in the water. It is also suspected that the ice is thinner over the bogs due to heat from the bogs. This would lead to better access to the terrestrial surroundings.

Over the long term, utilization and preference of woody species (considering availability by taxon, stem diameter and distance from the waterway) by beavers will be assessed on the permanent vegetation plots mentioned in Section 3.2.

3.4. Population Dynamics

Colony population counts by age class (kits, subadults and adults) will be obtained during October and November of each year when cache construction is underway and animals are easily observed (Easter-Pilcher 1990). Cache length and width (used to estimate cache area) will be measured at each colony site in late November. Cache area will be correlated with beaver numbers to examine the efficacy of cache area as a population index for beavers in the VKNP. Colony counts will be done each year as possible. Beaver colony counts will not only provide population dynamics information, but will also allow for analyses of the effect of beaver numbers upon the surrounding stream and habitat characteristics, including plant community diversity.

3.5. Distribution of Beavers

Remote sensing and geographic information system (GIS) technology will be utilized to monitor changes in beaver distribution across the landscape by locating beaver impoundments. Utilization of this technology is discussed in Section 4.2.

4. LANDSCAPE CHANGES

4.1. Changes in Vegetation

The permanent plots mentioned in Section 3.2. will be used in subsequent years to evaluate changes in woody and herbaceous vegetation species richness and abundance, changes in standing deadwood, and changes in diameter distribution. Some changes will be direct, due to cutting of stems by beavers, and some will be indirect, due to removal of competing stems and changes in water level. We will be able to evaluate whether changes in woody stems are from cutting of stems or indirect causes.

We will use one transect at each sample location to characterize changes in vegetative zones of dominance from lake/stream edge toward the forest interior. We will determine dominant and subdominant herbaceous and overstory species every 3 meters from the edge of the lake/stream to 33 meters inland. We hope to detect to some degree the distance from the original stream or lake bank that is influenced by beaver-induced changes in moisture and light.

4.2. Changes in Stream Characteristics and Water Distribution

Stream and lake characteristics will be measured adjacent to the mid-point on the center bank plot. These measurements (after Easter-Pilcher 1987) will include: stream gradient, velocity, width and depth, vertical water fluctuation, meander, slope distance between high and low water marks, bank height, slope, substrate and aspect, presence of a confluence, and understory woody taxa (number of stems available). Baseline measurements have been taken in July and August of 1997 and 1998, and additional measurements will be taken in subsequent years. This will allow assessment of changes in the characteristics of the waterway and surrounding habitat.

Remote sensing and geographic information system (GIS) technology will be utilized to monitor changes in beaver distribution across the landscape by locating beaver impoundments, and to monitor distribution of water across the landscape. GIS has been previously used to analyze long-term landscape alteration by beavers and was found to be an "indispensable tool" for studying the effects of beavers on boreal forest landscapes (Johnston and Naiman 1990 and Johnston *et al.* 1994). MacMillan, Furley, and Healey (1994) found that GIS capabilities were "essential" in analyzing data for use with hydrological models of surface water.

We will use the best resolution of satellite data we can afford in order to determine the distribution of water across the preserve landscape at baseline (1997 or the closest year when data is available) and periodically thereafter. Existing data is not liberally available because of low demand in this region and the number of cloudy/hazy days. Russian 5 m resolution color imagery from 1993 and SPOT 10 m panchromatic data from 1995 is known to be available. Groundtruth work has been proceeding in July and August of 1997 and 1998. Accuracy of water classification will be estimated through randomly located pixels in test fields. As a standard for comparison between years, future satellite data collection will be scheduled at times when water levels are expected to be near the annual low point. Water level gauge readings will be recorded at the time of satellite data collection. Utilizing GIS capabilities, these images will be compared through time to determine where beaver dams are being built and the change in surface water distribution. After several years of data classification, we will be able to use the information to build a model to predict future distribution and potential effects of beavers in other introduction scenarios at other preserves. Additionally, we hope to increase satellite image resolution when 1 meter resolution data becomes commercially available (predicted to be available in 1998).

4.3. Changes in Bird Biodiversity

Hair *et al.* (1979) documented significantly higher numbers of bird species in beaver impoundments. We expect structural heterogeneity to increase under the influence of beaver and that this heterogeneity will provide more microhabitats. It is reasonable to believe that bird biodiversity will increase as beavers spread their influence. Although we have had difficulty with initiation so far, we plan to monitor changes in bird diversity with permanent fixed-radius plots centered over the vegetation plots. Baseline work will begin in spring of 1999. Providing controls for long-term population changes that are not due to beaver-induced habitat changes will be the second challenge. We plan to pair beaver-occupied sites with unoccupied sites to test whether bird biodiversity has changed more positively in the occupied sites. We will use pre-occupation vegetation conditions to pair sites that are vegetatively most-similar.

5. EROSION PROCESSES ON ADJACENT AGRICULTURAL LANDS

Our second tactic is to learn more about the sources of sediment so that we can encourage agricultural practices that would slow the erosion. Even if beaver ponds cover the two river plains, they will have a limit to their sediment-holding capacity, so a long-term strategy must be established. We have begun working on this problem with Dr. Nikoli Torsuev and plan to have the sediment sources identified by the year 2000 so we may target some changes to agricultural practices soon afterwards. Points for measuring sediment loads and bank degradation have been established along the Sumka River on the preserve and upstream, and measurements will begin this fall.

6. THE FUTURE

6.1. Improvement to the Preserve and Surrounding Landscape

It is our expectation that the increased water levels from beaver ponds will have a long-term influence on the water table that will reverse the xerification of the landscape and the associated decline of mesic (especially boreal) species and loss of biodiversity. We also expect less sediment in lakes and bogs due to their entrapment by beaver ponds and the reduction of erosion from agricultural lands by improved agricultural practices. The life of these aquatic habitats will be extended by beavers and greatly extended by reduction of erosion. Another benefit, if we are very effective, will be the maintenance of agricultural productivity, since productivity of these lands adjacent to the preserve can only be perpetuated if there is little loss of topsoil. We expect greater diversity in plant and animal species as the more common, tall forest vegetation and common, xeric environments are replaced along portions of streamsides and lakeshores by saplings, beaver ponds, and more extensive mesic soil conditions. These are the conditions that must have prevailed when beaver populations were thriving here over 200 years ago.

6.2. Application of New Knowledge to Other Ecological Restoration Efforts

We expect results from our study to be applied to other areas. Our results will have predictive value for beaver habitat selection and could be used in the development of random-walk models. The greatest value for broader application will be the predictive value for habitat and biodiversity changes in other ecological restoration projects.

7. CONCLUSION

It seems that for most species, when a population is established, this population may fluctuate about a carrying capacity until the habitat changes. Animals typically cope through genotypic and phenotypic changes, by having a broad ecological amplitude, by moving on to better places. If the population is unable to cope with the environmental changes it will disappear. A rare form of coping is for a population to modify its habitat to suit its needs. This is exemplified best by two species: man and beaver. Let's hope the beaver works more quickly.

ACKNOWLEDGMENTS

Dr. Tatyana Gorshkova, Kazan Institute of Biology, Russian Academy of Sciences, was essential in the first year's efforts, and in fact, brought the ecological collaborators together initially. She used her vacation to translate the first season, and being a scientist herself, stimulated germinal discussions. Dmitry Gorshkov, Kazan State University (KSU) student and WMC student research aide, has done the greatest proportion of fieldwork with his sampling on the permanent vegetation plots and a year of counting beaver cuts, and has been our principal translator in 1998. Joe Nardinger, WMC student research aide, worked on vegetation plots and performed the essential task of tracking collections of unknown plants to ensure their identities were later entered on field forms. Evgenii Prokhorov and Andrei Pudovkin, VKNP staff, directed the trapping efforts. Nikoli Vasin, veterinarian from Kazan, performed the surgeries for the abdominal transmitters implants. Oleg Bakin, VKNP Vice-Director of Research, has helped with the orientation of U.S. scientists to the ecology of the preserve, plant identification, and supervision of some data collection.

We owe a special thanks to our financial contributors. In the United States, we have been supported by the Charles A. and Anne Morrow Lindbergh Foundation and the Montana Space Grant Consortium. In the Russian Federation, we have been supported by a grant from the Global Environment Facility and locally by the Tatarstan Ministry of Nature Conservation and Natural Resources and the administration of the Zelenodolsk District. These grants have made it possible to collect a large proportion of the baseline data without which, this project would not be possible.

REFERENCES

Collins, T. C., 1976, Population characteristics and habitat relationships of beaver (*Castor canadensis*) in northwest Wyoming, PhD thesis, Univ. of Wyoming, Laramie.

Davis, J. R., Von Recum, A. F., Smith, D. D., and Guynn, D. C., Jr., 1984, Implantable telemetry in beaver, *Wild. Soc. Bull.* **12**:322–324.

Easter-Pilcher, A. L., 1987, Forage utilization, habitat selection and population indices of beaver in northwestern Montana, Masters thesis, University of Montana, Missoula.

Easter-Pilcher, A. L., 1990, Cache size as an index to beaver colony size in northwestern Montana, *Wildlife Society Bulletin* **18(2)**:110–113.

Emlen, J. M., 1973, Ecology: an evolutionary approach, Addison-Wesley Publishing Co., Inc., Reading, Massachusetts.

Gauthreaux, S., 1978, The organization of forest avian communities, Proc. of the Workshop on Management of Southern Forests for Nongame Birds. U.S.D.A. Forest Service General Tech. Rep. SE-14. 17–37.

Hair, J. D., Hepp, G. T., Luckett, L. M., Reese, K. P., and Woodward. D. K., 1979, Beaver pond ecosystems and their relationships to multi-use natural resources management, U.S.D.A. Forest Service General Tech. Rep. WO-12.

Hartman, G., 1994, Long-term population development of a reintroduced beaver (*Castor fiber*) population in Sweden, *Conservation Biology* **8(3)**:713–717.

Johnston, C. A. and Naiman, R. J., 1987, Boundary dynamics at the aquatic-terrestrial interface: The influence of beaver and geomorphology, *Landscape Ecology* **1(1)**:47–57.

Johnston, C. A. and Naiman, R. J., 1990. The use of a GIS to analyze long-term landscape alteration by beaver, *Landscape Ecology* **4(1)**:5–19.

Johnston, C. A., Pastor, J., and Naiman, R. J., 1994, Effects of beaver and moose on boreal forest landscapes, in: *Landscape Ecology and Geographic Information Systems,* (R. Haines-Young, D. R. Green, and S. Cousins, eds.), pp. 237–254, Taylor and Francis, Inc., Bristol.

Karr, J. R. and Roth, R. R., 1971, Vegetation structure and avian diversity in several New World areas, *American Naturalist* **105(945)**:423–435.

Lavrov, L. S., 1934, The research of Voronezhskii National Preserve, Voronezh Publishing Co., Voronezh, U.S.S.R.

Lizarralde, M. S., 1993, Current status of the introduced beaver (*Castor canadensis*) population in Tierra del Fuego, Argentina, *Ambio* **22(6)**:351–358.

MacMillan, R. A., Furley, , P. A., and Healey, R. G., 1994, Using hydrological models and geographic information systems to assist with the management of surface water in agricultural landscapes, in: *Landscape Ecology and Geographic Information Systems*, (R. Haines-Young, D. R. Green, and S. Cousins, eds.), pp. 181–210, Taylor and Francis, Inc., Bristol.

McDowell , D. M., and Naiman, R.J., 1986, Structure and function of a benthic invertebrate community as influenced by beaver, *Oecologia* (Berlin) **68**:481–489.

Melquist, W. E., and Hornocker, M. G., 1979, Methods and techniques for studying and censusing river otter populations, Technical Report 8, Forest, Wildlife and Range Experiment Station, University of Idaho.

Naiman, R.J., and Melillo, J. M., 1984, Nitrogen budget of a subarctic stream altered by beaver (*Castor canadensis*), *Oecologia* **62:**150–155.

Naiman, R. J., Melillo, J. M., and Hobbie, J. E., 1986, Ecosystem alteration of boreal forest streams by beaver (Castor canadensis), *Ecology*, **67(5):**1254–1269.

Naiman, R. J., Johnston, C. A., and Kelley, J. C., 1988, Alteration of North American streams by beaver, *BioScience* **38:**753–762.

Pulliam, H. R., 1988, Sources, sinks, and population regulation, *American Naturalist* **132:**652–661.

Rogers, D. T., 1967, Bird populations of four different plant communities in the Blue Ridge Mountains, *J. Ala. Acad. of Science* **38(2):**172–176.

Schlosser, I., 1995, Dispersal, boundary processes, and trophic-level interactions in streams adjacent to beaver ponds, *Ecology* **76(3):**908–925.

Sielfeld, W., and Veneges, C., 1981, Poblamiento e impacto ambiental de *Castor canadensis* Khull en la Isla Navarino, Chile, *An. Inst. Patagonia* **11:**247–257 (In Spanish).

Willson, M. F., 1974, Avian community organization and habitat structure, *Ecology* **55:**1017–1029.

12

BEAVERS IN BRITAIN

Planning Reintroduction

David W. Macdonald and Fran H. Tattersall

Wildlife Conservation Research Unit, Department of Zoology
South Parks Road, Oxford OX1 3PS, United Kingdom

1. INTRODUCTION

1.1. Bringing Beavers Back to Britain

Over the last twenty years there has been regular discussion and debate about the desirability and feasibility of reintroducing the beaver to Britain (Yalden, 1986; Sage, 1981; Lever, 1980a, b, 1987, 1994; Halley, 1995; Macdonald, Tattersall, Brown, and Balharry, 1995). The European beaver, *Castor fiber*, once occurred in wooded areas throughout much of England, Wales and Scotland (Rackham, 1986; Clutton-Brock, 1991; Abyes and Yalden, 1995), but finally died out, in Scotland, in the 16th century (Conroy and Kitchener, 1996). The first positive step towards the species' return was taken in 1995, when Scottish Natural Heritage (SNH), the government agency responsible for nature conservation in Scotland, publicly committed themselves to investigating the possibility of reintroducing beavers to Scotland, and to coordinating any resulting releases.

There are sound conservation and legal reasons for taking this first step. Reintroductions are generally attractive in principle (Yalden, 1986) and provide a system for ensuring the conservation of species through increasing their range, minimizing the chances of extinction through local catastrophic change. The European beaver is still classified as Vulnerable by the IUCN/SSC Rodent Specialist Group (Amori and Zima, 1993), and a reintroduction to Britain would further expand its range and help ensure the species' long-term survival. In particular, beavers released in the UK would not be threatened by competition from North American beavers, *Castor canadensis*, which are present in Continental Europe. Reintroduction, especially of a keystone species such as the beaver (Macdonald *et al.*, 1995), also has distinct ecological advantages through the knock-on effects on ecosystem functioning, recreating conditions and opportunities lost to other species.

Beaver Protection, Managment, and Utilization in Europe and North America, edited by Busher and Dzięciołowski.
Kluwer Academic / Plenum Publishers, New York, 1999.

As a member of the European Community, Britain has a responsibility to investigate the feasibility of reintroducing 'Annex IV' species such as the beaver (Article 22, EC Habitats and Species Directive, EC 92/43); these are native species extirpated through man's activities in historical times, unable to naturally recolonise their former range. Under Article 11 (2) of the Convention of European Wildlife and Natural Habitats (Berne, 1979) the UK government is bound to encourage reintroductions of native species where this would contribute to their conservation. Finally, The UK Biodiversity Action Plan (HMSO, 1994) written in response to the 1992 Rio Convention on Biodiversity, of which the UK is a signatory, aims to conserve and enhance biological diversity in the UK. Where practicable, it aims to enhance populations and natural ranges of native species.

In this chapter we first explore key issues which need to be considered before assessing whether beavers are a suitable species for reintroduction to Britain, with particular emphasis on Scotland. We then document the early stages of the Scottish reintroduction and outline our vision of how a wider British reintroduction should proceed in accordance with IUCN guidelines (IUCN, 1998).

1.2. Past European Beaver Reintroductions

From past experience in Europe, the prognosis for successful establishment of a population of beavers in Britain is good. Since the 1920s European beavers have been reintroduced with considerable success in many countries in Europe and the former Soviet Union (Macdonald *et al.*, 1995; Nolet and Rosell, 1998). Macdonald *et al.* (1995) estimated that 53% of 87 reintroductions for which population estimates were available after at least five years could be considered successful, in that numbers have increased through reproduction.

Among previous European reintroductions, slow population growth or failure to establish was generally the result of release into unsuitable sites (Macdonald *et al.*, 1995), and most mortality among released beavers was caused by humans, for example through road accidents, hunting or entanglement in fish nets (Estève, 1988; Stocker, 1985). Some failures in Sweden and Switzerland were attributed to small propagule size, and this may also have affected reintroductions in Latvia and Finland. In countries such as Sweden and Finland, where small, very widely scattered pairs were released into large areas, it has taken about 40 years for the population to reach the stage where it requires control (Lavsund, 1977a, b). By comparison, in the former Soviet Union large groups of animals (50–100) were released, and required control after only 20 years (Zarkov, 1969, quoted by Zurowski and Kasperczyk, 1988). Although these comparisons are crude, deliberate establishment of a metapopulation by releasing larger groups, or small groups close together connected by habitat corridors, may speed up natural colonisation over a large area and reduce the risks posed by local extinctions.

Once they begin to breed, reintroduced populations generally increase at rates of between 20% and 34% annually (Zurowski, 1988; Migot and Rouland, 1989; Heidecke, 1983), although there may be periods of irruption and decline (Hartman, 1994). More established populations have slower growth rates; populations established in the 1940s in Poland, for example, increased at mean annual rates of 0–15% between 1980 and 1988 (Zurowski, 1988). Densities of reintroduced beaver are very variable, being affected by differences in habitat quality, management practices, length of time after reintroduction, and whether linear or total area are considered. In Sweden, overall densities in two regions were 0.06 and 0.10 colonies/km^2 in 1976, and 0.17 and 0.20 colonies/km^2 in 1986 (Hartman, 1994). In Poland, in the Suwalki Lakeland, density was 0.15 colonies/km^2 in 1985 (Zurowski and Kasperczyk, 1986).

There is no evidence that the beavers' origins affects the success of their reintroduction. Translocations of wild-caught animals are generally more successful (75%) than those of captive bred stock (38%; Griffith, Johnston, Scott, Carpenter, and Reed, 1989). In Poland third and fourth generation farmed beavers were able to build dams and lodges and collect food when released into the wild. They had a slightly higher litter size than wild animals, but were more susceptible to poaching because they were tamer (Zurowski and Kasperczyk, 1988). In the former GDR wild-caught animals adapted to release quicker than zoo-bred animals (Heidecke, 1985).

Many reintroductions, such as those in France and Switzerland, were notable for their lack of monitoring, strategic planning and standardised procedures (Migot and Rouland, 1989; Stocker, 1985). Thus far only the Netherlands reintroduction programme has had a well-structured approach, including a four-year public consultation and fact-finding period, a five-year experimental phase with an extensive associated research programme, and an assessment of population viability and overall project success (Nolet, 1994; Nolet and Baveco, 1996).

1.3. The Scottish Reintroduction: Current Status

Within Britain, beavers persisted in Scotland until the 16th century (Conroy and Kitchener, 1996), long after they became extinct in England and Wales. Scotland remains the least populated country within the United Kingdom, with probably the largest areas of wilderness and is therefore most suited to a beaver reintroduction. SNH, Scotland's government conservation agency, undertook in 1995 to investigate the feasibility and desirability of reintroducing beavers to Scotland, and for carrying out the restoration if appropriate.

Following IUCN guidelines (see below), SNH initially identified three stages in their work programme: first, to investigate the ecological feasibility of reintroducing the beaver; second to investigate its desirability; and third to establish a reintroduction protocol and programme (Cooper, 1996). As part of their feasibility study, SNH commissioned several research projects, investigating the history of the beaver in Scotland (Conroy and Kitchener, 1996), potential costs of a reintroduction (Tattersall and Macdonald, 1995), current habitat availability (Webb, French and Flitsch, 1997), a method of assessing site suitability (Macdonald, Maitland, Rao, Strachan, and Tattersall, 1997; Macdonald, Tattersall, Rushton, Rao, Maitland, and Strachan, subm.), interactions with fisheries (Collen, 1997), and impact on hydrology (Gurnell, 1997).

The desirability of reintroducing beaver to Scotland was recently assessed by SNH through a broad-ranging public consultation, as required under Article 22(a) of the Habitats Directive. The consultation document (SNH, 1998) reported on SNH's initial feasibility studies and solicited the views of both the general public and of informed or interested parties, such as foresters and farmers. The consultation specifically asked whether people supported the proposal to restore the beaver to Scotland, and if so, whether a reintroduction should begin by 2001.

Whether or not the proposal to reintroduce the beaver to Scotland goes ahead will depend to a large part on the results of the consultation. If the restoration is considered desirable, SNH will proceed to apply to Government for a licence to carry out the project, and will begin to implement the third stage of planning a release protocol and carrying out the reintroduction programme.

2. PLANNING A BRITISH REINTRODUCTION: A NEW MODEL?

Clearly the reintroduction of a species as large and potentially controversial as the beaver, into a modern, largely urban society such as Britain's, must be approached extremely carefully and in accordance with IUCN (1998) guidelines. We hope that the British reintroduction of the beaver will serve as a model for future species restoration projects. As the 21st century looms, the *ad hoc* past releases of beavers are no longer appropriate, either morally, legally, or scientifically.

Morally, at the simplest level it is undesirable to reintroduce an animal without some assessment of current habitat suitability. It is wasteful of resources to reintroduce species without assessing the potential viability of the population and the potential for population growth. It is also unethical to release species without some assessment, firstly, of whether they have the necessary resources for survival, and secondly of the potential for reintroduced animals to expand into areas where they could cause damage, resulting in a wider undesirable economic impact.

Legally, any proposed reintroduction of beavers to Britain would have to be approved and licensed by the Government's Department of the Environment, Transport and the Regions, because domestic legislation makes it illegal to release into the wild any animal which is not normally resident (Wildlife and Countryside Act, 1981). Implicit in this would be approval by Conservation Agencies such as English Nature, Scottish Natural Heritage, Countryside Council for Wales and the Joint Nature Conservation Council. As a related issue, beavers would currently not receive legal protection, as they are not normally resident in the UK; this would need to be rectified under domestic legislation. On an international level, the EU 'Habitats Directive' requires "proper consultation with the public concerned" prior to any reintroductions. Questions of acceptability and economic damage are of paramount importance.

Finally, a modern reintroduction should make full use of the modelling, computing and animal-tracking technology now available to assess habitat suitability, population growth and spread, before, during and after releases. Recent developments in modern computing and modeling techniques mean that it is now feasible to carry out a 'desktop survey' of an entire country to focus on potentially useful sites, and to gauge their connectedness, although the suitability of such 'long-listed' sites must be validated in the field. Further, we can now estimate how long a sub- or metapopulation is likely to survive in a particular area or network of sites. It is within our grasp, therefore, to select a network of sites to which we can attach a given probability that released animals will persist through time.

2.1. IUCN Guidelines

Prior to any releases in Scotland or elsewhere in Britain, it is essential to evaluate formally beaver reintroduction in accordance with IUCN guidelines (IUCN, 1998). These guidelines are intended to *"help ensure that reintroductions achieve their intended conservation benefit and do not cause adverse side-effects of greater impact"*. They list criteria to be considered including the availability of suitable habitats and animals for release, potential socio-economic and environmental impacts, and the ease with which populations and damage to property could be controlled. As well as a large number of pre-project studies, IUCN guidelines call for a planning, preparation and release phase, and a follow-up and assessment phase (Table 1). The viability of the project should be reassessed at each stage, and should be aborted if it appears to be unfeasible or undesirable.

2.2. The Role of Simulation Modeling in Reintroductions

Modern simulation modeling techniques have the potential to play a significant role at every stage of a reintroduction, from an initial assessment of habitat availability, to

Table 1. Pertinent stages and activities in the reintroduction of the beaver to Britain following IUCN (1998) guidelines

1. Pre-project activities: assessing biological feasibility
   Study previous reintroductions determine the beaver's critical needs assess the impact of the beaver on its environment assess habitat availability and consider habitat restoration assess likelihood of establishing a minimum viable population establish criteria for selecting release sites assess likely impact of previous causes of decline assess availability of animals to be released.
2. Pre-project activities: assessing socio-economic and legal desirability
   Assess likely long-term financial and political support assess impacts, costs and benefits to local people investigate means of reducing damage to property assess attitudes of local communities through public consultation check local, national and international legislation acquire relevant permits.
3. Planning, preparation and release
   Obtain approval of relevant bodies secure funding for all stages construct a multi-disciplinary team of experts identify short- and long-term indicators of success design a monitoring programme establish policies on interventions determine the release strategy select final release sites mount a public education/awareness campaign arrange capture, travel and quarantine and screen animals for disease
4. Post-release activites:
   Implement the monitoring programme carry out associated studies on released animals collect and investigate mortalities assess population viability assess the need for further extensive releases evaluate project success and cost-effectivenes evaluate socio-economic and environmental impact continue habitat protection or restoration as necessary intervene, revise or abandon project as necessary disseminate information about the project.

planing releases and predicting population spread. Incorporation of field data gathered though monitoring initial releases allows refinement of models and subsequent management actions (Nolet and Baveco, 1996). Modeling can also be used to assist in the identification of 'success indicators' (IUCN, 1998), for example, a natality rate that generates a prediction of a certain population persistence or a threshold population size needed to reduce the predicted risk of extinction to below a pre-defined level. This follows the adaptive management approach to reintroductions advocated by Sarrazin and Barbault (1996).

The two major techniques which have a role in reintroductions are Geographical Information Systems (GIS) and Population Viability Analysis (PVA) GIS provide technologies for the storage and analysis of cartographic data and have been used extensively in assessing the suitability of landscapes for different species (Aspinall and Veitch, 1993). While it is possible to identify potential sites for reintroducing species, this does not guarantee that a reintroduction will succeed. Population modeling offers a route to predicting the likely consequences of reintroduction schemes (e.g. Howells and Edwards-Jones, 1997). The combined use of GIS and PVA may be particularly valuable as a tool for evaluating management options or release strategies (Macdonald, Mace, and Rushton, 1998; Macdonald *et al.*, subm., 1997). Nonetheless, models in wildlife management should not be taken as representations of the truth but should be used as problem-solving tools (Starfield, 1997).

2.2.1. Assessing Habitat Availability Using GIS. Assessing landscapes for suitable reintroduction sites is potentially a very costly exercise, particularly for species such as beavers, which have large home ranges and can disperse widely. One obvious approach, which may circumvent the need for extensive field visits, is to use cartographic data stored and analysed on a GIS.

While GIS may offer a fast route for analysing large geographical data sets, their application in identifying potential sites for reintroducing species is dependent on suitable data being available and a suitable associative link between the data and the ecological requirements of the species concerned. Remotely-sensed imagery from aerial photographs and satellite imagery have been used extensively in assessing habitat suitability for many animal species (further discussion in Macdonald *et al.*, 1998) and these are particularly useful where species have very restricted habitat requirements which are easily defined in terms of the mapped information.

Even where a species has an easily definable and recognizable habitat, such as wooded river bank for beaver, the accuracy with which suitable sites are identified will depend on the quality and quantity of landscape data available. For example, in addition to the presence of woodland on a riverbank, hydrological and soil data are likely to be important to assess fully sites for beavers. The addition of each habitat criterion would tighten the constraints determining site suitability, and reduce the area of potential release sites. At the purely technical level, data quality may impact on the accuracy with which suitable habitats are recognized in the GIS. For example, land cover features and river data might be identified from different maps, which may not be oriented with the same accuracy, or collected at the same scales. Use of a relatively coarse scale, and large errors in overlap, would tend to overestimate habitat availability.

2.2.2. Assessing Population Viability Using PVA. While an assessment of the physical structure of the release site is an essential prerequisite of any reintroduction scheme, it is obvious that some assessment of the likely success of the species following release at potential sites is also desirable. It is generally out of the question to tackle this through experimentation. The only approach is predictive modeling, which has its own limitations.

PVA models the effects on populations of many interacting variables such as random variation in population size, loss of genetic variation and catastrophic events such as fire and flood. For a given set of characteristics, population viability analysis (PVA) estimates the expected time to extinction, or the chances of a population becoming extinct within a specified time (Boyce, 1992). Nolet and Baveco (1996) used PVA to evaluate the release strategy for beavers in the Netherlands, and to gauge the population's chances of survival. A population's viability can be assessed using simulation modeling, and off the shelf computer programmes such as VORTEX (Lacy, 1993) have recently become available to do this.

However, different PVA programmes may yield different results and they should be interpreted with caution (e.g. Macdonald *et al.*, 1998; see also below).

2.2.3. PVA and GIS as Management Tools. Population models are most useful to reintroduction schemes when they are combined with GIS. These 'process-based models' rely on the premise that the distribution of a species in the landscape arises from interactions between individual behavioural processes such as home-range behaviour, territoriality and dispersal, and the life-history processes of births and deaths. In these models the habitat data act as templates on which the population processes occur and the distribution of organisms among habitats emerges as the model is run. These approaches are inevitably much more complex than straightforward habitat availability estimates, but have been used to predict the distributions of other species with complex rapidly changing distribution patterns (e.g. red and grey squirrels; Rushton *et al.*, 1997).

While process-based models may be highly suited to analyzing the spread of species such as the beaver, the accuracy of the predictions will be determined primarily by the simplifications and assumptions made in the model. These can be grouped into two: those associated with the identification of suitable habitats for occupation, and those associated with modeling the life-history processes of birth, deaths and dispersal. The effects of these on the validity of the model's predictions need to be assessed critically. Problems relating to habitat identification have already been discussed above.

The second major source of potential inaccuracies in the model relate to life history processes. These models require detailed information on mortality, reproduction and so on, at each stage. Life-history data from newly established populations in similar habitat to the release sites may be more appropriate than information derived from research on established populations because they represent conditions experienced at reintroduction. A potentially more important problem concerning the impacts of variations in the values of individual life-history variables are the values used to parameterise dispersal. Dispersal is one of the most under-studied processes in population dynamics, simply because it is usually very difficult to identify whence animals have dispersed. Furthermore, the factors which initiate dispersal are strongly linked to social-behavior and these are not processes that are easily modeled in the form of equations. The processes initiating dispersal may, therefore, be grossly over-simplified.

3. THE FIRST STAGES: SHOULD WE REINTRODUCE THE BEAVER TO BRITAIN?

The pre-project stage of the potential reintroduction of the beaver to Britain began with Macdonald *et al.*'s (1995) review of its feasibility and desirability, and their conclusion that the idea merited further attention. This was followed by SNH's decision to investigate the possibility of reintroducing the beaver to Scotland, their commissioning of further research (Collen, 1997; Conroy and Kitchener, 1996; Gurnell, 1997; Macdonald *et al.*, subm., 1997; Webb *et al.*, 1997) and, most recently, a widescale public consultation exercise (SNH, 1998). In this and the following section we briefly review the biological feasibility and the socio-economic and legal desirability of reintroducing beavers to Britain, and to Scotland in particular.

3.1. Is It Biologically Feasible to Reintroduce the Beaver to Britain?

We have already discussed previous reintroductions above (see also Macdonald *et al.*, 1995; Nolet and Rosell, 1998), and conclude that these have generally been successful. As a result of these reintroductions, wild-caught animals are available for reintroduction from populations in Scandinavia, Germany, Poland and eastern Europe, though we do not yet know which subspecies was present in Britain. The beaver's critical needs and their environmental impact are also well known (reviewed in Macdonald *et al.*, 1995) and do not need to be detailed here. In brief, beavers are aquatic and require year-round access to water fringed with herbaceous and woody vegetation. They have a major impact on their environment through felling trees and building dams, creating ponds of water. In Britain, Coles and Orme (1983) suggest that during the Mesolithic beavers had at least as great an effect on the landscape as humans. Mesolithic human settlements near water may have taken advantage of areas already cleared by beavers, which have previously been misinterpreted as the result of human activity. Beavers may also have aided the spread of Alder, *Alnus glutinosa*, through the British Isles during the Mesolithic by creating suitable habitats for its establishment (Chambers and Price, 1985; Chambers and Elliott, 1989). Beavers reintroduced into Britain are eventually likely to increase vegetation diversity and create and maintain wetland habitats with consequent benefits for aquatic fauna.

Hunting and habitat loss were probably the main causes of extinction of the beaver (Conroy and Kitchener, 1996). As there is no longer a tradition of hunting beaver, this is unlikely to be an important cause for concern in modern-day Britain, though killing of perceived 'nuisance' beavers might be more problematic. Habitat loss is potentially more important, and assessment of habitat availability and the likelihood of establishing a minimum viable population are critical to assessing the feasibility of reintroducing the beaver. Two stages are crucial to this assessment - first an assessment of the overall availability of habitat potentially suitable for occupation by a MVP of beavers, and secondly selection of the release sites themselves.

3.1.1. Is Sufficient Habitat Available to Support a Minimum Viable Population? The British landscape of today is radically different from that of 400–1000 years ago, so an assessment of habitat availability is vital. Scotland is probably the best area in Britain in which to concentrate the search for suitable sites, as it still has remote, isolated areas, which have been less affected by man than elsewhere.

The habitat size required to maintain a viable population will, clearly, depend on carrying capacity and estimated minimum viable population size (MVP). There is no consensus in the literature about either the MVP for beavers, nor the habitat size required to maintain it. Estimates of the total length of river needed to maintain a viable population vary from 6km (Stocker, 1985) to 50km (Heidecke, 1986; M. Stubbe, pers. comm.), while in marshes the estimated minimum area required by one family is 10ha (Ouderaa, Fey, Stelt, and Zadelhoff, 1985).

Small populations, such as those of newly reintroduced animals, are inherently more at risk from demographic and environmental stochasticity than larger populations (Caughley, 1994). Both processes have been important in preventing or slowing establishment of released populations of beavers: in Sweden small, widely separated, releases made it difficult for animals to find mates (Hartman, 1994), and in Poland seasonal meltwater floods prevented beavers from establishing in mountainous sites (Zurowski, 1987). Small populations are also susceptible to loss of heterozygosity and fitness. However, there is no evidence for inbreeding depression in the Swedish beaver population, which has extremely low levels of genetic variability at DNA fingerprinting loci (10.8–23.6%), and is monomorphic at major histocompatibility complex class 1 and 2 loci (Ellegren, Hartman, Johansson and Anderson, 1993).

Qualitative assessment suggest that in Scotland at least three river systems might be suitable: the Spey River system north of Newtonmore and south of Grantown ; Strath Dee from Braemer to Banchory; and the Tay river system including the main Tummel and Garry tributaries, north of Perth and especially north of Dunkeld. These three areas are heavily wooded, drain the Cairngorm mountains, and would probably be in genetic contact with one another.

Studies by Webb *et al.* (1997) and Macdonald *et al.* (subm., 1997) used GIS to provide a more formal assessment of Scotland. Macdonald *et al.* (subm., 1997) used a GIS to identify habitat suitable for beaver home ranges and habitat through which animals could disperse. They assumed that the minimum area capable of supporting beavers was either a contiguous block of woodland along more than 3.1km of river bank, and/or blocks of marshland of greater than 10ha. Dispersal habitats were assumed to be areas of non-montane land through which a river corridor or stream system ran. Since the land surface of Scotland is extensively permeated by water courses the areas of non-dispersal, non-home range habitat were confined to mountainous areas. Only 94 sites were identified as suitable and only three of these (Loch Insh, Glen Meinich and Glen Esk) had areas of contigu-

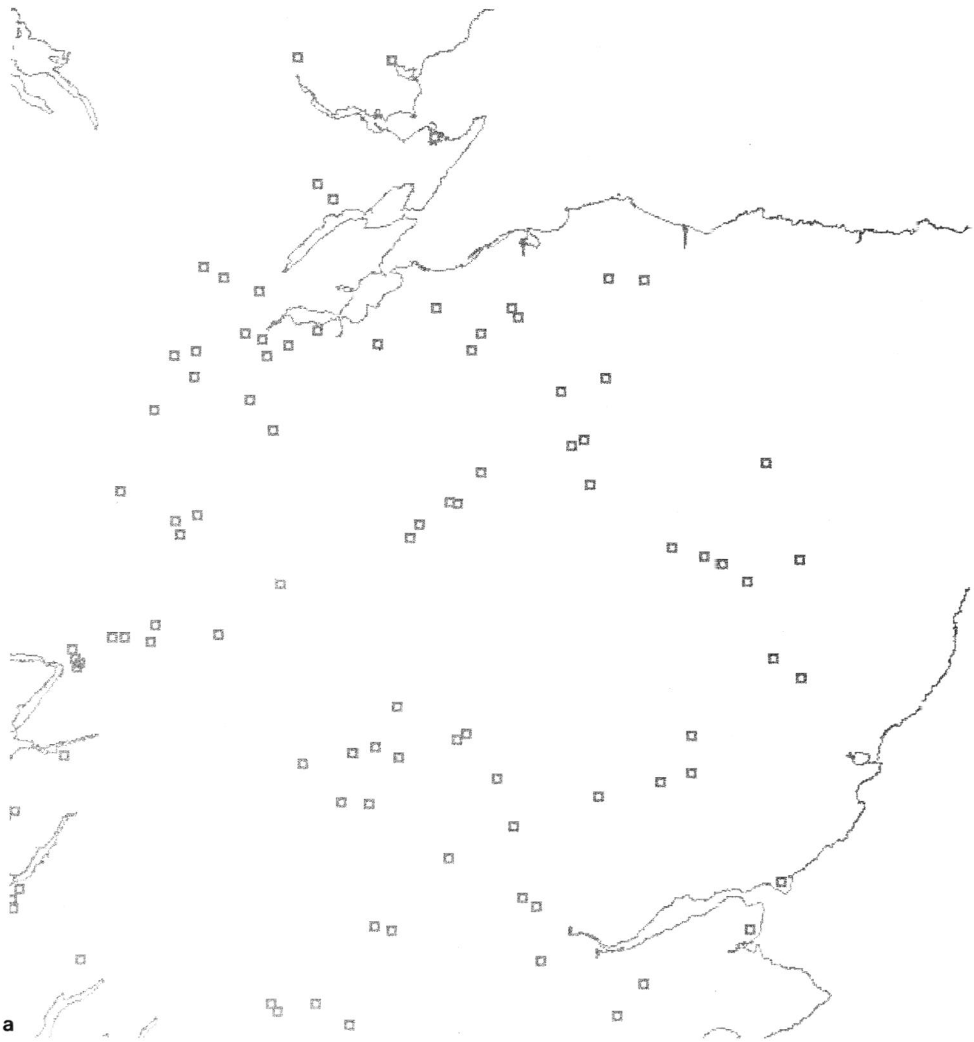

a

Figure 1. The predicted distribution of sites suitable for a) (this page) occupation by beaver colonies and b) (next page) dispersal by beavers, in the study area (four 100 km National Grid Squares: NH, NJ, NN and NO) in Scotland.

ous habitat that could potentially support 10 or more beaver families. Sites were widely separated and few large sites were within close proximity of each other (Figure 1a). Webb *et al.'s* (1997) analysis suggested that suitable areas were concentrated in a central block around the major river catchments of the Dee, Spey, Ness, Tay and Lomond. They estimated that, even without habitat restoration, the present landscape of Scotland could support 200–1000 beavers. Both studies indicate that the landscape in the region is well connected (Figure 1b), suggesting that there is considerable potential for beaver dispersal.

Thus far there has been no formal assessment of habitat in Wales. In England, Perkins (unpub. report) carried out field surveys to assess the potential of the Norfolk Broads, a series of man made lakes created through centuries of extensive peat cutting,

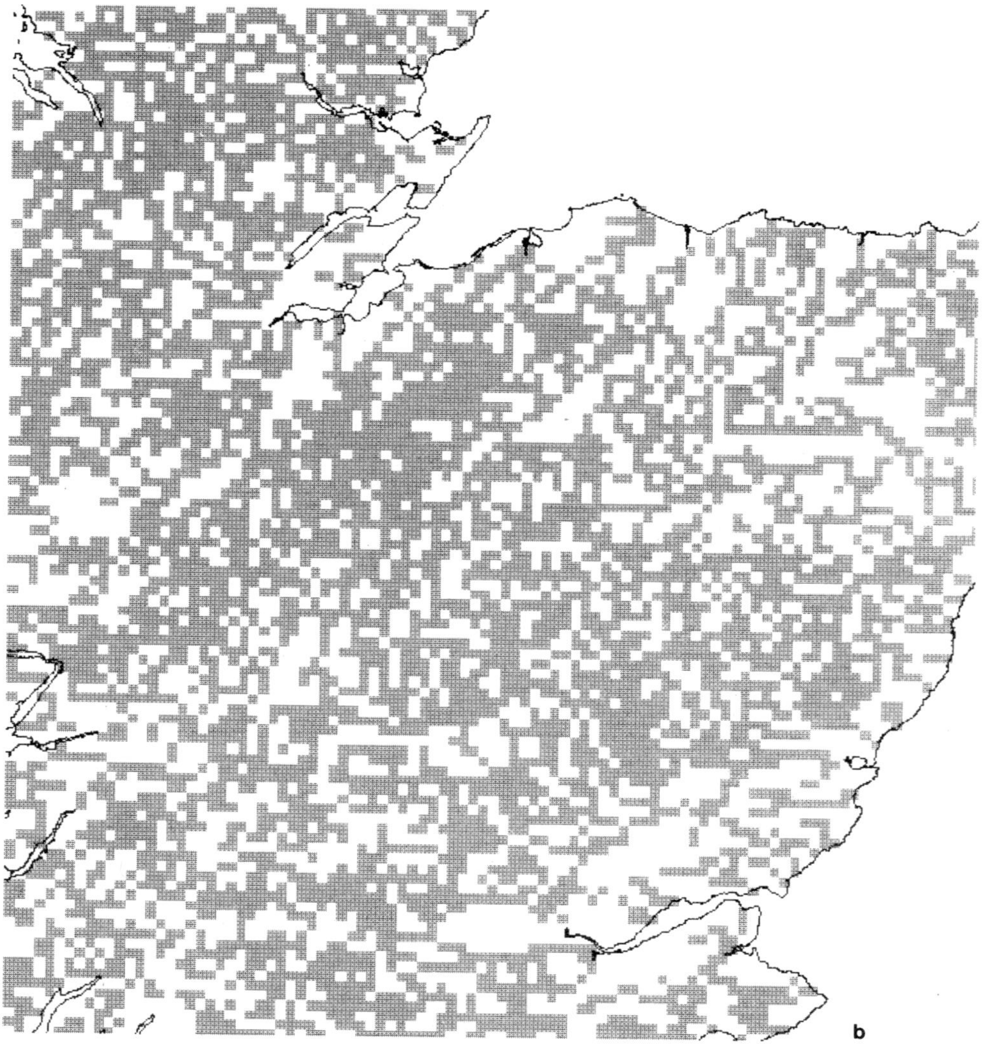

Figure 1. (*Continued*)

and now one of Europe's most important lowland wetlands. They surveyed over 137 km of waterway following criteria given in Macdonald *et al.* (subm., 1997), and provisionally estimated that the area could support at least 24 families in two large but separate areas. However, the area also contains a dense network of dykes and ditches which were not included in the analysis, which may provide suitable habitat or access to it. Further investigation of habitat suitability and connectivity through use of GIS modelling is underway.

3.1.2. Identifying and Assessing Suitable Release Sites. Although beavers can live in a wide range of habitats, release sites should be carefully chosen to maximise the chances of establishment and breeding, and minimise mortality and conflict with humans. Beavers should only be reintroduced into areas where water is present year-round at a depth of at least 1m, particularly close to potential lodge or burrow entrances. Although beavers can survive in water depths of as little as 50cm (Ouderaa *et al.*, 1985), they are more likely to build dams in shallow water. Similarly, they are less likely to build dams on large rivers, but can live on water courses less than 2m wide. A range of softwoods, ideally aspen and willow, or herbaceous vegetation must be available throughout the year, preferably close to the lodge or burrow. Release sites should be part of a network of suitable areas within a river system, to ensure that as the population expands it is able to colonise a wide area. Ideally, sites should be within the historical range (Griffith *et al.*, 1989), and should be in protected areas such as Nature Reserves.

Our approach in identifying potential sites consists of four phases combining GIS-based analyses of habitat data with field survey and population modelling (Macdonald *et al.*, subm., 1997). Each phase identifies a list of potential sites which is progressively reduced in subsequent phases as more fine-scale assessments of suitability for release are considered. The final stage includes the time dimension, assessing the potential long-term suitability of a restricted range of sites for population viability (this is explored in the next section). The four Phases are as follows:

Phase 1. The national-scale identification of potential sites for releasing beavers and dispersal routes in Scotland, using mapped habitat information held in a GIS.

Phase 2. The design and field testing of a preliminary survey protocol and scoring system for site selection or rejection, using a sub-sample of sites identified in Phase 1.

Phase 3. Detailed site surveys and habitat mapping of the best sites identified in Phase 2 to provide a subjective assessment of the receiving potential of those sites for a release group of beavers.

Phase 4. The development of a simulation model to explore the potential spread of an introduced beaver population from potential release sites identified in Phase 3 (see next section).

Macdonald *et al.* (subm., 1997) carried out such an assessment for Scotland. At a subsample of nine sites identified in Phase 1, they assessed habitat during a single, brief field survey. The physical parameters of the watercourse and its bank, availability of bankside vegetation, human disturbance and adjacent land use were measured and graded into categories according to their impact on released beavers (Table 2a and b). None of the sites emerged as ideal for beavers - of a potential score of 30, the highest score was 17.4 for site 8. For the three most promising sites (2, 7 and 8, Table 3) carrying capacity was estimated assuming a minimum 3.1km of suitable bank-side habitat per beaver colony of two adults and two to three subadults (Nolet and Baveco 1996) (Phase 3). To calculate the amount of suitable habitat available a detailed habitat survey protocol (National Rivers Authority, 1995), supplemented by existing data, was used to create an accurate vegetation map for land within 100m of the water. This is the maximum distance beavers are thought to travel regularly from the bank in search of food (Stocker 1985; Heidecke 1989). Carrying capacities were 6–8 colonies for site 8, 2–4 colonies for site 7 and 2–3 colonies for site 2.

The inherent limitations of spatial resolution and quality in the GIS data mean that a field survey will always be a vital component of any system for selecting release sites. In order to be useful, a field survey must allow comparison between different types of habitat (in our case marsh, loch and river), and should provide an objective, quantitative means of summarizing site suitability. In order to assess as many sites as possible, it should also be quick to complete and require little specialized knowledge, at least for a preliminary assessment. A weighted scoring system fulfills these requirements.

Clearly this exploration has no pretensions to completeness: its goal is to describe an approach rather than to decide upon a site. In doing so, however, we have uncovered inter-

Table 2. Criteria for scoring potential sites*

a)		Score x	Score 0	Score 1	Score 2
Water	mean depth (m) across watercourse	<0.5 (still water only)	<1, >6	1-2, 4-6	2-4
	width (m)	<0.075	<2, >300	2-10, 100-300	10-100
	flow speed (m/s)	>1	0.6-1	0.3-0.6	0.3-0
	fluctuation level (m)	>3, rapid	2-3	1.5-2	<1.5
	pollution		inorganic/high	organic/low	none
Bank	substrate	rock/gravel	sand/rock	podsol	peat/loam
	angle (°)	>90	80-90	60-80	<60
	height (m)		<0.5	0.5-1.5	>1.5
Vegetation	emergent/in channel cover (%)		<40	40-60	>60
	herbaceous cover within 100m (%)	0-20	20-40	40-60	>60
	shrub/hardwood cover within 20m (%)	0-20	20-40	40-60	>60
	shrub/hardwood cover 20-100m (%)		<40	40-60	>60
	average hardwood diam (cm)		>20	10-20	<10
Human	disturbance/mortality	Certain	High (>8)	Medium (4-8)	Low (<4)
	potential for damage	Certain	High (4-8)	Medium (1-4)	Low (0)

b)	Certain	0 points	1 point	2 points
Disturbance/ mortality	road (B or larger) <100m	powered boats	walkers/cyclists	protected area
	railway line <100m	commercial fishing	anglers	no or restricted public access
	intensive net/trap fishing <3km	hunting with hounds	rowing boat	
		shooting	keepering	
		built up area <1km	no formal protection	
		development possible	built up area 1-5km	
Potential for damage	hardwood plantation <100m and little alternative food	farmed land < 500m	500-3km	>3km
	road (B or larger) <100m	road/railway <500m	500-3km	>3km
	railway line <100m	plantations <500m	500-3km	>3km
	any intensively used land <50m	buildings <500m	500-3km	>3km

* a) At each candidate release site 5–11 point sampling positions (selected at roughly 1km intervals) were assessed for factors relating to the water, bank, vegetation and human impact. At each point, each factor was given a score of x, 0, 1 or 2, and the scores were totaled. Where a criteria ranked X, the score for that point was 0, regardless of the quality of other factors. The mean total for all sampling points provided the overall score for each site. b) Human impact was assessed using a points system based on either the presence or absence of particular features, and/or their proximity. The presence of features likely to cause disturbance/mortality to beavers, or to be damaged by them merited 0 points, while those unlikely to disturb or be damaged merited 2 points; intermediate factors merited 1 point. A low total score indicates high potential for damage or disturbance, and therefore scores 0; a large point total indicates low potential for damage or disturbance and therefore scores 2. The presence of one or more features certain to cause beaver mortality, or to be damaged by beavers, scored an X , and the entire sampling point returned a 0 score.

Table 3. Scores for habitat suitability for beaver.
Scores are mean total scores for a number of sampling points (n)

Site	1	2	3	4	5	6	7	8	9
Mean score	12.4	15.9	6.0	8.5	2.0	6.0	16.5	17.4	14.3
n	5	9	6	6	5	9	11	7	6
Rank	5	3	7.5	6	9	7.5	2	1	4

esting questions. For example, the best knowledge available within SNH identified a list of nine candidate sites which overlapped rather little with those selected by GIS in Phase 1, and it will be important to explore what divergence of criteria cause this distinction between naturalistic expertise and desktop quantification.

The most obvious conclusion from the field surveys is that few, if any, of the nine sites considered would actually be capable of supporting beavers, either because of an absence of food or because of a combination of unfavourable physical characteristics, or because of intensive adjacent land-use.

3.2. Is It Socio-Economically and Legally Desirable to Reintroduce Beavers?

Neither domestic nor international legislation present a barrier to a beaver reintroduction in Britain; indeed legislation positively encourages the British government to at least investigate the possibility, and to follow this through should it be feasible and desirable. With a large species such as the beaver, which has the potential to alter vastly its ecosystem, the main questions are whether a reintroduced population will cause unsustainable economic damage, and whether the local human population will be in favour of the beavers.

3.2.1. What Are the Likely Impacts, Costs and Benefits to Local People and How Can Damage to Property Be Reduced? Beavers reintroduced to Britain have the potential to cause damage to forestry, crops, agricultural land and river banks through their feeding, burrowing and damming, but also have positive, though less tangible, benefits such as flood control, water conservation and renewal, and stabilisation of groundwater levels (Table 4). The socio-economic impacts of European beavers, and their control, was reviewed in detail by Macdonald *et al.* (1995). There is no evidence that beavers negatively impact on hydroelectric schemes and they have little impact on conifer plantations. While beavers can cause erosion of river banks, they can also prevent it by slowing water flow (Naiman, Melillo, and Hobbie, 1986). Where beavers cause problems they can be humanely removed, and their dams can be destroyed or an overflow pipe inserted. That beavers in Britain might damage commercial timber and crops is true. However, once a population is established it might be possible to raise revenue by charging people to see beavers (as a precedent, the RSPB successfully charge visitors to their reserves), or even by hunting

Surprisingly, in Scotland, fishermen have been among the most vocal opponents of a beaver reintroduction. Collen (1997) consulted all 62 Scottish District Salmon Fishery Boards and 60 angling clubs in Scotland on the proposal to reintroduce beaver. Of the 59 organisations which responded, 70% were against the reintroduction and 19% felt they did not have enough information to comment. Many fishermen believe that beaver dams will impede migrating salmonids, and that an increase in coarse woody debris resulting from

Table 4. Positive and negative impacts of beavers

Positive	Negative
Improve water quality	Flood farmland, roads and woodland
Prevent flooding downstream	Feed on softwoods and crops
Regulate water flow	Burrow through dykes
Create and maintain wetlands	Erode river and ditch banks
Benefit other species	Undermine fields and roads
Raise the water table	
Conserve water	
Are a potential tourist attraction	

beaver activity would be detrimental to fishing interests; some also expressed concern that beaver would damage riparian woodland planting schemes designed to enhance food resources for salmonids. Collen (1997) concluded that fisheries authorities would be unlikely to support the beaver's restoration unless the population was actively managed.

The actual likelihood of beavers damaging the Scottish fishing industry is probably remote, but nonetheless possible. There is little information on the relationship between European beaver and fish, and reports of a conflict of interests comes only from the North American beaver, *Castor canadensis*, in North America. In Europe, reintroduced beaver coexist with salmon, trout, pike and eels, and are generally thought to have a beneficial impact on fish stocks, by increasing habitat and invertebrate food (Collen, 1997). However, localised problems may occur when trout spawning areas are damaged by siltation, or, in hot weather, when the water in a beaver pond heats up sufficiently to deplete oxygen levels.

We do not anticipate that beavers will cause serious damage in Scotland, and where they do cause damage we believe they can readily be controlled and their impact minimised. The impact of any damage might be more important at a local level, and therefore wherever possible the potential impact of beavers should be assessed at a local level, and a management plan should be devised to cope with local-scale problems.

3.2.1. What Are the Attitudes of Local People? Following a number of scientific studies, in 1998 SNH undertook a public consultation exercise to determine whether or not there was the public will to proceed with restoring the beaver to Scotland. The consultation document (SNH, 1998) summarised the results of their research, assessed a potential reintroduction against IUCN guidelines, and invited comment from interested parties and the general public. Specifically, two questions were asked: "Do you support a proposal to restore the European beaver to the wild in Scotland?", and "If so, do you support such a reintroduction beginning during the next three years?". In addition to the formal consultation document, which was widely distributed among relevant professionals and available to the public from SNH, an Internet site was devoted to the issue, and questionnaire leaflets were available. The consultation is now closed, and SNH are analysing the results.

3.2.3. Long-Term Financial and Political Support. If the reintroduction of the European beaver is to require medium-term habitat restoration or long-term wardening then this will involve costs. Some of these might be met by government through the statutory nature conservation agencies, or relevant ministries. Wildlife charities are another obvious source of funds. However, we think some attention might usefully be given to a less orthodox source of finance: mitigation swapping. This idea, which descends from the concept of carbon offset schemes, is outlined by Macdonald, Rushton, and Mace (in press). In brief, a substantial injection of funds into conservation could come from industry in exchange for unavoidable environmental damage.

Clearly, whatever the source of the funds, likely long-term costs of beaver reintroduction should be considered in advance and budgeted for. Our suspicion is that it might be some years (perhaps 5 to 10) following a successful introduction before an expanding population began to generate reports of nuisance, and eventualities on this time-scale and longer should be anticipated.

3.3. Conclusions

Suitable sites are potentially available for beaver reintroductions in Britain. Scotland still has high quality bodies of water in areas with low human population density, in which suitable sites might be found. Thus, in view of past European successes, and the large body of literature and practical expertise that has accumulated, reintroduction of the beaver to Britain is certainly feasible.

Is it desirable, however? North American beavers can certainly cause extensive damage, but the effects of European beavers are less well documented. Ultimately it will be too complex to predict accurately the amount of damage that beavers are likely to cause, though we can be sure they will cause some. Sufficient political and public goodwill must be present either to pay compensation or absorb the costs of this damage. However, with careful selection of release site, taking into account likely migration within and between water courses, it should be possible to reduce the likelihood of damage occurring, at least while the population remains small and localised.

4. PREPARATION, PLANNING, AND RELEASE

4.1. Preparation and Planning

We have already touched on key elements of the "business plan" that would be needed in a preparatory step to beaver reintroduction. Approval of the relevant bodies is essential, as is securing funding for all stages of the programme. We also think that the inter-disciplinarity of the team is important: relevant issues include biology, agriculture, forestry and fisheries, among others. Our own team often refers to what we call the Conservation Quartet: four ingredients of any successful project. These are research, education, community involvement and implementation. All four elements are essential, and in this context the importance of a public awareness campaign should not be underestimated.

4.2. Determining the Release Strategy and Selecting Sites

Site selection and determination of the release strategy should be done in parallel using a combination of GIS and PVA, each depending on the other. At each of a sub-sample of the best potential areas identified by GIS, a PVA should be run utilising the characteristics of that particular site. Clearly the number of animals a site can accommodate will depend on its carrying capacity, which will in turn depend on variables such as its size, food availability and hydrology. Whether or not one can create a viable metapopulation will depend, in turn, on the juxtaposition of sites, their connectivity and overall carrying capacity. Two examples of how one might go about determining a site-specific release strategy already exist in the literature.

4.2.1. VORTEX Simulations. Using the available information on life history characteristics and population ecology of beavers, Macdonald *et al.* (1995) ran VORTEX simulations to assess the time to extinction for a beaver population. In this case the model assumed no inbreeding depression; monogamous, density dependent reproduction from three to twelve years; an even sex ratio; a maximum litter size of six; 70% of males bred; maximum breeding age was 12 years. Beavers were assumed to have a cyclical effect on their habitat: habitat quality diminished by 2% per annum for two years and then recovered at 2% per annum for the next 10 years (these figures were arbitrarily chosen for demonstration purposes only). A loss of heterozygosity of 1% per generation (Soulé, 1986), and the risk that the population might become extinct within 200 years were considered tolerable. For each assessment 100 simulations were performed over 100 years, with a reporting interval of 10 years.

Three sets of simulations were performed (Figure 2). The first (Figure 2a) assumed an area with a mean ± SD carrying capacity of 400 ± 80 beavers (or one family of four per kilometre) in which there was a single release of 5, 10, 20, 30, or 50 pairs. Releases involving 20–50 pairs had a greater than 80% chance of survival over 100 years, while a release of five pairs had only a 40% survival rate over 100 years. The second and third simulations (Figures 2b and 2c) assumed a mean ± SD carrying capacity of 200 ± 40 on 50 km of river with and without supplementary reintroductions of two pairs every second year for the first 20 years. Where the reintroductions were supplemented there was a large improvement in the chances of survival for all populations: except for a release of 30 pairs, all supplemented reintroductions had a greater than 80% chance of survival after 100 years, while unsupplemented, only releases of 50 pairs had greater than 80% survival chance, and releases of 5 pairs had a less than 30% survival chance.

These preliminary results, particularly the comparison between supplemented and unsupplemented releases, demonstrate the usefulness of simulation modelling, which should play an important part in formulating a reintroduction strategy for beavers in Britain. In particular, the effects of creating a metapopulation by reintroducing beavers at several different sites, should be explored. Nolet and Rosell (1994) suggested that territorial behaviour might play a greater role in limiting density in sequentially released populations than in single releases because early arrivals claimed larger territories than later ones; where possible such behavioural factors should be included in the model.

4.2.2. Individual-Based PVA. Macdonald *et al.* (subm.) ran an individual-based population dynamics module which simulated life histories of individual beavers and their dispersal within the GIS-held landscape to predict how beavers released into three sites would survive and disperse. The model was run under two dispersal and two carrying capacity scenarios, and estimated the mean number of subpopulations in each year, and the mean total metapopulation size in each year, following release of adults at each site. The mean number of subpopulations developing under each scenario at each site were in all cases relatively low; at no time did the number of subpopulations exceed ten. The most obvious feature was the low level of colonization away from all of the release sites. No metapopulations went extinct under any of the dispersal or carrying capacity scenarios at site 8. At sites 2 and 7, metapopulations went extinct in 3–30% of all runs. In this example, carrying capacity and dispersal distance were varied, but one could equally vary, for example, numbers of animals, sex and age ratios, or compare sequential versus one-off releases at a particular site.

a

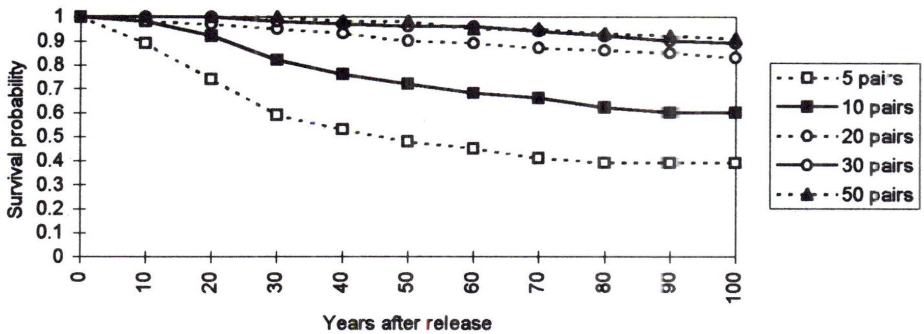

b

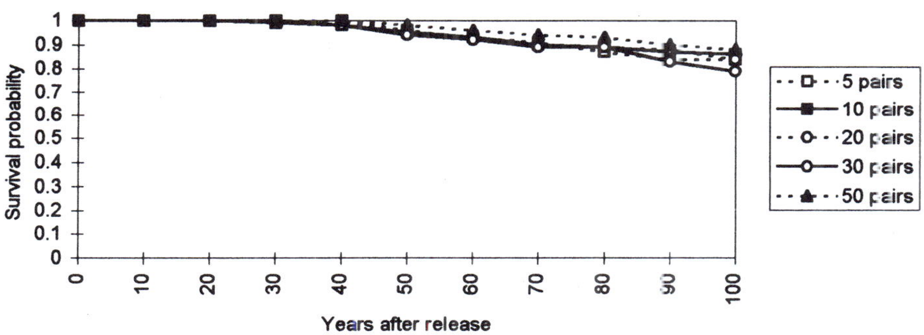

c

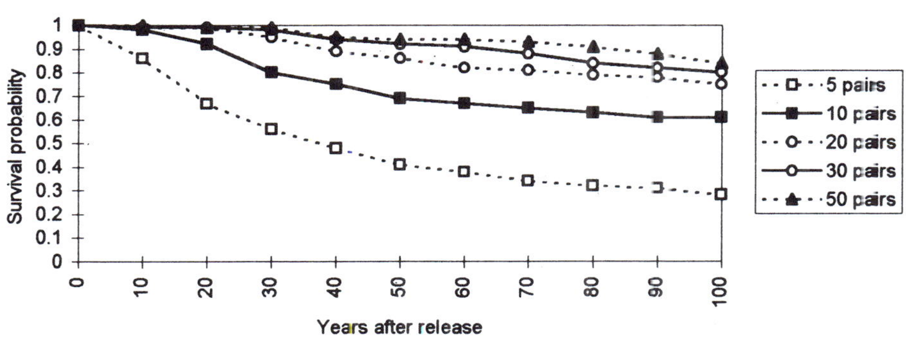

Figure 2. Analysis of beaver population viability using VORTEX simulations. a) The model assumes an area with a mean ± SD carrying capacity of 400 ± 80 beavers (or one family of four per kilometre) in which there is a single release of 5, 10, 20, 30, or 50 beaver pairs. For each assessment we performed 100 simulations over 100 years, with a reporting interval of 10 years. b) The model assumes a mean ± SD carrying capacity of 200 ± 40 on 50 km of river in which there is a release of 5, 10, 20, 30, or 50 beaver pairs, followed by supplementary reintroductions of two pairs every second year for the first 20 years. For each assessment we performed 100 simulations over 100 years, with a reporting interval of 10 years. c) The model assumes a mean ± SD carrying capacity of 200 ± 40 on 50 km of river in which there is a single release of 5, 10, 20, 30, or 50 beaver pairs, with no supplementary reintroductions. For each assessment we performed 100 simulations over 100 years, with a reporting interval of 10 years.

What do the results of the modeling tell us about the likely future of the beaver following reintroduction? The results suggest that spread of the beaver following reintroduction at the selected sites is likely to be slow and relatively small scale. Of the three sites, release at sites 8 and 2 gave rise to similar metapopulation sizes. These results are not surprising given the close proximity of the two release sites. The slower expansion from site 2 relative to site 8 reflects the smaller initial starter population at site 2 prior to invasion of the larger adjacent site 8. Site 7 appeared to give rise to a much lower metapopulation size, with slower growth rate. Indeed the size and number of subpopulations predicted to develop away from site 7 were so low as to render this non-viable as a release site. The main reasons for this would appear to be related to abundance of sufficient habitat to allow the establishment of subpopulations.

It is worth noting that release into unsuitable sites and small release populations at release sites have been considered to be important factors determining failure in previous attempts at reintroducing the beaver (Macdonald *et al.*, 1995). The low densities and low number of sites colonized suggest that introduced beavers would be susceptible to catastrophic events. While the potential effects of such events were not investigated formally here (because there is no meaningful way of modeling such events) it is easy to predict their consequences (see, for example, Macdonald *et al.*, 1998). Given the low numbers of subpopulations and their small size, it is likely that the beaver metapopulations would not be viable over the long-term at any of the release sites if there was any likelihood of catastrophic mortality arising from factors such as flooding or road kills. Small populations are also, of course, at risk from inbreeding effects. We are mindful of the hazards of setting too much store by the outputs of one particular model, and alert to the fact that, for example, different PVA programmes may yield different results (e.g. Macdonald *et al.*, 1998). To emphasize this caution, we have explored alternative modelling approaches to test the sensitivity of the results presented here (South *et al.*, in press).

These results all suggest that to secure the long-term viability of reintroduced beavers in Scotland is likely to require habitat creation. In the meantime, it is likely that reintroduced beavers would have to be the subject of metapopulation management. However, that conclusion does not nullify the conservation or aesthetic desirability of attempting to reintroduce beavers. Indeed, the conservation of many species, including for example, many bird species in the UK, is largely dependent on sustained intervention, and around the world acceptance of the need for active metapopulation management is increasingly widespread (e.g. African wild dog, *Lycaon pictus*: Mills, Ellis, Woodroffe, Maddock, Stander, Rasmussen, Pole, Fletcher, Bruford, Wildt, Macdonald, and Seal, 1998).

4.2.3. Habitat Modification. Habitat improvement and creation schemes could also be initiated to provide suitable habitat to aid the spread of beavers. Such schemes have been successfully employed for otter reintroductions (Andrews, Barratt, Liles, Crawford, Scott, and Lenton, 1986), and generally improve the wildlife value of sites. Effective management of food resources for beavers is particularly important in small habitats, which are vulnerable to over-exploitation. Some sites may need to be restocked with fast growing willow species (Stocker, 1985). Willows and shrubs should be planted close to the river edge to provide food and shelter. Well established willows should be cut back in groups approximately every 10 years, to encourage new growth (Stocker, 1985). Grass in meadows near rivers or streams should not be cut, to encourage diversity of herbaceous species.

4.3. Monitoring

Monitoring of released animals, and their impact on the environment, is an essential part of any reintroduction. Monitoring is vital not only to assess the effectiveness of the release strategy (for example, whether animals survive and breed, whether subpopulations come into contact with one another), but also to assess the biological and socio-economic impact of the released animals. Monitoring should be designed to provide information to improve PVA model predictions, and to improve our understanding of how to manage the population. Thus monitoring might provide information on, for instance, litter size, longevity, and dispersal, which allow further refinement of future release strategies.

Monitoring should be carried out at every stage of the reintroductions, but should be most intensive at the beginning of the project, when only a small number of animals are present. Radio-tracking beavers after release is facilitated by their tendency to remain close to rivers and to leave obvious feeding signs. Radio-collars have been used in America (Busher, Warner, and Jenkins, 1983; Buech, 1985), but are often lost; radio-harnesses are more successful (Gillespie, 1976). Radio-transmitters can be implanted (Daview, Von Recum, Smith, and Guynn, 1984; Nolet, 1994), and in Britain this method would require a Home Office licence. Subcutaneous transponders (PIT tags) should be implanted for individual identification.

4.3.1. Designing a Monitoring Programme. Throughout Europe there is a spate of interest in monitoring wildlife. This is doubtless fuelled by the generally disheartening realisation that many things are changing. In the UK, Macdonald *et al.* (1998) have summarised the arguments in favour of systematic monitoring of mammals. Clearly, it would be imperative to monitor carefully the fate of reintroduced beavers. The principles underlying this exercise do not differ from those established for other British mammals, and neither is it likely to be difficult in practice. Obviously, the monitoring programme should be planned in advance of the reintroduction. It should involve the collection of data not only on numbers of beavers, but also on their habitat and impact.

4.3.2. Indicators of Success. We suggest that indicators of the success of beaver reintroduction could usefully embrace several different spheres of judgement. In addition to the obvious criteria of whether they become established in satisfactory numbers, it will be informative to record also measures of environmental impact, nuisance, eco-tourism, educational penetration and public perception. Wherever possible, these indices should be quantified and enumerated for statistical analysis. In the case of biological measures, indices should be relevant to demography (e.g. breeding success, longevity) and metapopulation processes.

4.3.3. Establishing Policies on Interventions. Fail-safe mechanisms should be in place to mitigate any disadvantageous turn of events in the reintroduction programme. One reason why we are optimistic about the idea of reintroducing beavers in the UK is that they are an easy species to manage. This is evidenced by the ease with which they have been widely exterminated. In the event that some individuals cause problems it is likely that they could easily be captured or killed. It is important that the public is introduced to the notion of such intervention so that the possible local need to manage beavers is not perceived as contradicting a general desire to foster their presence. In this context, we are mindful of the possibility that relatively straightforward conservation management of problematic beavers might not be seen as straightforward in terms of animal welfare. Adequate monitoring data, and refined models, should facilitate the task of predicting and pre-empting problems.

4.4. Release

Currently, the UK has particular quarantine laws. These are under review, but irrespective of changes, it will be vital to consider carefully any veterinary implications regarding the beavers to be imported for release.

4.4.1. Veterinary Considerations. The only diseases that beavers might transmit to birds and mammals in Britain, which are not already present, are rabies and tularaemia. These should be preventable by statutory quarantine procedures and prophylactic treatment for tularaemia. European beavers are less susceptible to tularaemia than North American beavers (Stocker, 1985), and there have been no reports of epidemics affecting European beavers at the population level (Romasov, 1992). Beavers also suffer from pseudotuberculosis and pneumonia, and the parasitic gut fluke *Stichorchis subtriquetrus* (100% of beavers in Lithuania and 84.4% in Russia) and gut nematode *Travassosius rufus* (45% and 63–100%: Mickus, 1995; Tyurnin, 1984). A programme of health screening and dosing with anthelminthic would be advisable prior to the release of beavers for reintroduction. Beavers are also associated with a specialised ectoparasite, the Beaver beetle *Platypsyllus castoris* (Stocker, 1985; Besuchet, 1978; Buchholx and Sikora, 1984) but this does not affect any other European mammals, apart from the introduced North American beaver.

If released beavers were able to spread into agricultural land, fencing or other protective measures might need to be taken. Local farmers (if there were any) would probably want assurances that nuisance beavers could be trapped. In the Netherlands, it was decided that the SBB, the organisation which released the animals, would pay compensation (B. Nolet, pers. comm.), and this option needs to be considered. If the site is in a public area, information boards could be on display, and measures might need to be taken to restrict access close to artificial or natural burrows or lodges, at least until the population became established. Potential benefits of beavers to local people, such as eco-tourism, should be investigated and brought to fruition as quickly as possible.

5. A NEW MEMBER OF THE BRITISH FAUNA? HOPES AND FEARS

5.1. Follow-Up and Assessment

The necessary follow-up will come under all of the sub-headings mentioned in this chapter. For example, it will be important to:

a. Implement the monitoring programme.
b. Carry out associated studies on released animals.
c. Collect and investigate mortalities.
d. Assess population viability.
e. Assess the need for further, extensive releases.
f. Evaluate project success and cost-effectiveness.
g. Evaluate socio-economic and environmental impact.
h. Continue habitat protection or restoration as necessary.
i. Intervene, revise or abandon project as necessary.
j. Disseminate information about the project.

5.1.1. Determining the Success of the Project. Before the reintroduction a series of clear criteria and objectives should be defined, with which to gauge the success of the project. These objectives could consider breeding success, survival and causes of mortality, emigration from the release site, complaints of damage, evidence of beneficial effects of beavers, and their image among local people. So far as possible, objectives should be qualitative and easily measured, and should include both long-term and short-term goals. The results of the project should be assessed at regular, pre-set time periods in order to ensure that objectives are being met, or to consider the implications of not meeting objectives and intervene as necessary. Once a population has been successfully established in an area, the desirability of further, more extensive reintroductions or translocations, particularly into new water basins, should be assessed.

5.1.2. Long-Term Monitoring and Research. Long-term monitoring should involve annual estimations of the population size and distribution. Estimating beaver numbers can be done by estimating the number of families present in an area from the number of lodges or burrows (W. Zurowski, pers. comm.); these can sometimes be seen from the air (G. Hartman, pers. comm.). In Poland data come from hunters, farmers and foresters (B Kasperczyk, pers. comm.). In the former GDR monitoring was carried out by volunteers, each covering a small (10–20km^2) area (M. Stubbe, pers. comm.). The use of volunteers to monitor extensive releases should be considered as various British conservation organisations already have experience of using volunteers for large projects (e.g. the Mammal Society Wood Mouse *Apodemus sylvaticus* survey, Mallorie and Flowerdew, 1994). Detailed records should be kept of, for example, deaths, damage, movements, building and breeding, in order to assess the success of the project. DNA fingerprinting could be carried out on released animals to monitor relatedness in the developing population.

5.1.3. Use of Modelling Throughout the Reintroduction Process. Monitoring information gathered during initial releases should be used to refine model parameters. Model runs based on improved estimates could then be used to predict population development and direct subsequent management actions. Such an approach is being applied to the beaver reintroduction in the Netherlands (Nolet and Baveco, 1996). Modelling could also be used to assist in the identification of 'success indicators' (IUCN, 1998), these might take the form of a natality rate that generates a prediction of a certain population persistence or a threshold population size needed to reduce the predicted risk of extinction to below a pre-defined level. This follows the adaptive management approach to reintroductions advocated by Sarrazin and Barbault(1996).

5.2. Beavers in England and Wales

There is little doubt that within the UK, Scotland is the most obvious nation with which to begin exploring the topic of reintroduction. However, this should not distract conservation agencies in other countries from considering the possibilities. In particular, it may not be helpful that attention tends to have focused on an all-or-nothing approach to this issue, revolving around the question of whether a widespread, viable introduction is achievable. We see much of UK conservation as being inescapably interventionist. Localised and highly managed introductions of beavers could allow useful exploration of the issues, provide worthwhile experience and be positive in their own right. It is not necessary to await the long-term outcome of the Scottish experience before beginning to explore the possibilities elsewhere in the UK. Indeed, some areas of England may well be even more suited to beavers.

5.3. Managing Beavers in the UK—the Long-Term Future

The possibility of beavers becoming an emblem of wetland conservation reserves in the UK is an exciting one. There seems little doubt that such reserves could be created and fostered if the political commitment and funding were forthcoming (e.g. Macdonald *et al.*, in press). In particular, while we suspect that further analysis will reveal that reintroduced beavers in Scotland would require continued metapopulation management, we consider that this would be entirely in accord with the general necessity for interventionist conservation typical in the modified landscapes of western Europe.

While our models may be suited to analyzing the spread of the beaver, the species has not yet been introduced into Scotland and so clearly we cannot validate our predictions. Thus, while the results may suggest that the reintroduction of the beaver into the short-listed sites may meet with limited success, these conclusions need to be considered with care. The accuracy of the predictions will be determined primarily by the simplifications and assumptions made in the model. These can be grouped into two: those associated with the identification of suitable habitats for occupation by beaver in Scotland and those associated with modeling the life-history processes. The effects of these on the validity of the model's predictions need to be assessed critically.

The results from the GIS component suggest that large blocks of habitat are rare and sparsely distributed in Scotland, indeed the three sites potentially capable of supporting more than 10 beaver colonies are all in widely separated river catchments. Overall, the metapopulation densities predicted to be present after 50 years are low for all release sites. This in part reflects the fact that very little of the study area appears to be suitable for supporting beavers. Indeed the original desktop survey of the nine sites studied in detail suggested that none could support more than 11 colonies of beaver, while other estimates extrapolated from the River Corridor Survey data suggested a maximum of only 8 colonies.

Finally, we stress again two aspects of this issue which somewhat limit worries. First, as a matter of risk analysis, the likelihood of a reintroduction going irreversibly wrong would seem to be close to nil. Beavers are all too easy to trap and to kill. If some unforeseen circumstance caused their presence in the UK to become undesirable, they could rather easily be removed. Second, if the principle of interventionist management is accepted, then experimental releases could be explored while accepting that they did not necessarily constitute an irreversible commitment. There is thus scope for a tentative advance that may ease some political pressures on the debate.

5.4. Conclusions in a Wider Context

The IUCN Reintroduction Specialist Group listed the following as suitable objectives for a reintroduction project (IUCN, 1998): to enhance the long-term survival of a species; to re-establish a keystone species in an ecosystem; to maintain and/or enhance biodiversity; to provide long-term economic benefits to a local or national economy; and to promote conservation awareness. A reintroduction of beavers to Britain could fulfil most, if not all of these objectives, and meets UKCINC (1979) and IUCN criteria for reintroductions.

We believe that the reintroduction of beavers to Britain is desirable for the following reasons:

i. Beavers are a keystone species, which play a pivotal role in maintaining and regulating aquatic ecosystems (SBB, 1983; Balodis, 1994; Johnston, Pastor, and Naiman, 1993).
ii. As well as directly increasing faunal diversity by their very presence, beavers encourage biological diversity through their impact on their environment; in particular their presence could benefit otter and water Vole (Tyurnin, 1994; Balodis, 1984; Danilov, 1995).

iii. Despite the success of reintroductions in Europe, beavers are still classified as vulnerable by the IUCN/SSC Rodent Specialist Group (Amori and Zima, 1993). A reintroduction to Britain would further expand their range and help to ensure the species long-term survival.

iv. Inclusion of exercises such as the use of PVA and GIS to determine release strategy and site respectively, genetic monitoring to determine relatedness, and detailed assessment of the impact of beavers on their environment would add scientific interest to the project and this could set a precedent for reintroductions in general.

v. Because of their unique behaviour and appearance beavers are inherently interesting to humans, and could be a 'flagship' species for nature conservation, heightening people's awareness and understanding of wildlife in Britain. Beaver reintroductions might foster schemes for habitat improvement or creation, all of which would benefit other wildlife.

vi. There is philosophical and aesthetic value in returning a native species to the wild, particularly one as prominent and interesting as the beaver. Reintroducing beavers gives us the opportunity of "atoning for past follies and trying again to support them as permanent elements of the fauna" (Morris, 1986).

In summary, a review of the likely outcome of reintroducing beavers provides an appealing list of benefits. Risk assessment suggests that the costs may be few and rather easily compensated, managed or, if necessary, reversed. We are inclined, therefore, to support the proposal.

ACKNOWLEDGMENTS

We have greatly enjoyed working with SNH in exploring these issues, successively with David Balharry and Mairie Cooper. The current review was sponsored by the International Trust for Nature Conservation, who we warmly thank. Many of the ideas developed here were born of discussions with Steve Rushton and Andy South, whom we thank along with Georgina Mace and Lauren Harrington.

REFERENCES

Amori, G., and Zima, J., 1993, Threatened rodents in Europe: species status and some suggestions for conservation strategies, *Fol. Zool.* **43:**1–9.

Andrews, E., Barratt, M., Liles, G., Crawford, A., Scott, H., and Lenton, E., 1986, The Otter Haven Project, *Vincent Wildlife Trust Report* **1985:**5–15.

Aspinall, R., and Veitch, N., 1993, Habitat mapping from satellite imagery and wildlife survey data using a Bayesian modelling procedure in a GIS. *Photogrammetic Engineering and Remote Sensing* **59:**537–543.

Aybes, C., and Yalden, D.W., 1995, Place-name evidence for the former distribution and status of Wolves and Beavers in Britain. *Mamm. Rev.* **25:**201–227.

Balodis, M. M., 1994, Beaver population of Latvia: history, development and management, *Proceedings of the Latvian Academy of Sciences,B* **7/8:**122–127.

Besuchet, C., 1978, Reintroduction of *Platypsyllus castoris* Rits. (Coleoptera, Leptinidae) in Switzerland, *Bull. Soc. Entom. Suisse.* **51:**291–292. [In French, with English summary].

Boyce, M. S., 1992, Population viability analysis *Ann. Rev. Ecol. Syst.* **23:**481–506.

Buchholz, L., and Sikora, S., 1984, *Platsysyllus castoris*. Ritsema, 1869 (Coleoptera, Platypsyllidae), a new beetle for the fauna of Poland, *Przeglad. Zool.* **28:**501–506. [In Polish, with English summary].

Buech, R. R., 1985, Methodologies for observing Beavers (*Castor canadensis*) during the activity period, in: *Nocturnal mammals: techniques for study*, (R. P. Brooks, ed.), Pennsylvania State University, School for Resources, Research Paper 48.

Busher, P. E., Warner, R. J., and Jenkins, S. H., 1983, Population density, colony composition, and local movements in two Sierra Nevaden Beaver populations. *J. Mamm.* **64:**314–318.

Caughley, G., 1994, Directions in Conservation Biology. *J. Anim. Ecol.* **63:**215–244.

Chambers, F. M., and Elliott, L., 1989, Spread and expansion of *Alnus* Mill. in the British Isles: timing agencies and possible vectors, *J. Biogeog.* **16:**541–550.

Chambers, F. M., and Price, S. -M., 1985, Palaeoecology of *Alnus* (alder): early post-glacial rise in a valley mire, North-West Wales, *New Phytol.* **101:**370–400.

Clutton-Brock, J., 1991, Canadian Beaver, *Castor canadensis,* in: *The Handbook of British Mammals,* 3rd ed., (G. B. Corbet and S. Harris, eds.), p. 572, Blackwell Scientific Publications, Oxford.

Coles, J. M., and Orme, B. J., 1983, *Homo sapiens* or *Castor fiber*?, *Antiquity* **57:**95–102.

Collen, P., 1997, Review of the potential impact of reintroducing European Beaver *Castor fiber L.* on the ecology and movement of native fishes, and the likely implications for current angling practices in Scotland. Scottish Natural Heritage Review No. 86.

Conroy, J. W. H., and Kitchener, A. C., 1996, The European Beaver (*Castor fiber*) in Scotland: A review of the literature and historical evidence. Scottish Natural Heritage No. 49.

Cooper, M., 1996, Reintroduction of the European beaver to Scotland, *Reintroduction News* **12:**11–12.

Curry-Lindahl, K., 1967, The Beaver, *Castor fiber* Linnaeus, 1758 in Sweden - Extermination and Reappearance, *Acta Ther.* **12:**1–15.

Danilov, P. I., 1995, Canadian and European Beaver in the Russian Northwest - distribution, number and comparative ecology, in: *The Third Nordic Beaver Symposium, 1992, Finland*, pp.10–16, Finnish Game and Fisheries Research Institute, Helsinki.

Daview, J. R., Von Recum, A. F., Smith, D. D., and Guynn, D. K., 1984, Implantable telemetry in Beaver, *Wildl. Soc. Bull.* **12:**322–324.

Ellegren, H., Hartman, G., Johansson, M., and Andersson, L., 1993, Major histocompatibility complex monomorphism and low levels of DNA fingerprinting variability in a reintroduced and rapidly expanding population of beavers, *Proc. Natl. Acad. Sci. USA* **90:**8150–8153.

Estève, R., 1988, An analysis of Beaver mortality in Haute-Savoie, *Bievre* **9(2):**171–176. [In French, with English summary].

Gillespie, P. S., 1976, Summer activities, home range and habitat use of Beavers, M.Sc. thesis, University of Toronto.

Griffith, B., Johnston, C. A., Scott, J. M., Carpenter, J. W., and Reed, C., 1989, Translocation as a species conservation tool: status and strategy, *Science* **245:**477–480.

Gurnell, A. M., 1997, Analysis of the effects of Beaver dam-building activities on local hydrology. Scottish Natural Heritage Review No. 85.

Halley, D., 1995, The Proposed reintroduction of the beaver to Britain, *Reintroduction News* **10:**17–18.

Hartman, G., 1994, Long-term population development of a reintroduced beaver (*Castor fiber*) population in Sweden, *Cons. Biol.* **8:**713–717.

Heidecke, D., 1983, Bestandssituation des Elbebibers und Schutzmassnamen in der DDR, *Mitteilungen der BAG Artenschutz Magdeburg* **6(2):**1–8.

Heidecke, D., 1985, Erste Ergebnisse der Biberumsiedlungen in der DDR, *Zoologische Abhandlungen Staatliches Museum fur Tierkunde Dresden* **41(12):**137–142.

Heidecke, D., 1986, Bestandssituation und Schutz von *Castor fiber albicus, Zoologische Abhandlungen Staatliches Museum fur Tierkunde Dresden* **41(9):**111–119.

Heidecke, D., 1989, Okologische bewertung von Biberhabitaten, *Saugertierkundliche Mitteilungen* **3:**13–28.

HMSO, 1994, Biodiversity, The UK Action Plan, HMSO, London.

Howells, O., and Edward-Jones, G., 1997, A feasibility study of reintroducing wild boar *Sus scrofa* to Scotland: Are existing woodland large enough to support minimum viable populations. *Biol. Cons.* **81:**77–89.

IUCN, 1998, Guidelines for reintroductions, IUCN/SSC Reintroduction Specialist Group, IUCN, Gland, Switzerland and Cambridge, UK.

Johnston, C. A., Pastor, J., and Naiman, R. J., 1993, Effects of Beaver and Moose on boreal forest landscapes, in: *Landscape ecology and geographic information systems,* (R. Haines-Young, D. R. Green and S. Cousins, eds.), pp. 237–254, Taylor and Francis, London.

Lacy, R. C., 1993, VORTEX: a computer simulation model for population viability analysis, *Wildl. Res.* **20:**45–65.

Lavsund, S., 1977a, Beaver population dynamics in Sweden, in: *Proceedings from the Nordic Symposium on the beaver, Ramsele, 23rd-26th Sept. 1975, Institute of Forest Zoology, Research Notes 26,* (S. Lavsund, ed.), pp.26–28.

Lavsund, S., 1977b, Beaver hunting methods in Sweden, in: *Proceedings from the Nordic Symposium on the Beaver, Ramsele, 23rd-26th Sept., 1975, Institute of Forest Zoology, Research Notes 26,* (S. Lavsund, ed.), pp 79–81.

Lever, C., 1980a, No to the Beaver, *Ecos* **1**:22–23.

Lever, C., 1980b, No Beavers for Britain, *New Scientist* **80:**812–814.

Lever, C., 1987, The case against the Beaver, *Illustrated London News* **269 (April, 1987):**38.

Lever, C., 1994, The proposed reintroduction of the beaver to Britain, *Reintroduction News* **9:**14–15.

Lima, S. L., and Zollner, P. A., 1996, Towards a behavioural ecology of ecological landscapes, *TREE* **11:**131–135.

Macdonald, D. W., Tattersall, F. H., Brown, E. D., and Balharry, D., 1995, Reintroducing the European Beaver to Britain: nostalgic meddling or restoring boidiversity?, *Mamm. Rev.* **25:**161–200.

Macdonald, D. W., Maitland, P., Rao, S., Strachan, R., and Tattersall, F., 1997, Development of a protocol for identifying beaver release sites. Scottish Natural Heritage Research, Survey and Monitoring Report No. 93

Macdonald, D.W., Tattersall, F. H., Rushton, S., Rao, S., Maitland, P. and Strachan, R., subm., Reintroducing the Beaver (*Castor fiber*) to Scotland: A protocol for identifying and assessing suitable release sites. *Anim Cons.*

Macdonald, D. W., Mace, G., and Rushton, S., 1998, Proposals for future monitoring of British mammals, Department of the Environment, Transort and the Regions.

Macdonald, D. W., Rushton, S., Mace, G., in press, Mammal Conservation in Britain: is there a radical future?, in: Future priorities for the conservation of mammalian diversity, (A. Entwhistle and N. Dunstone, eds.). Cambridge University Press, Cambridge.

Mallorie, H. C., and Flowerdew, J. R., 1994, Woodland small mammal population ecology in Britain: a preliminary review of the Mammal Society survey of Wood Mice *Apodemus sylvaticus* and Bank Voles *Clethrionomys glareolus*, 1982–87, *Mamm. Rev.* **24:**1–15.

Mickus, A., 1995, The European Beaver (*Castor fiber* L.) in Lithuania, in: *The Third Nordic Beaver Symposium, 1992, Finland*, pp.44–45, Finnish Game and Fisheries Research Institute, Helsinki.

Migot, P., and Rouland, P., 1989, La reintroduction du castor en France: essai de synthese et reflexions, Bulletin Mensuel Office National de la Chasse **132:**35–43.

Mills, M. G. L., Ellis, S., Woodroffe, R., Maddock, A., Stander, P., Rasmussen, G., Pole, A., Fletcher, P., Bruford, M., Wildt, D., Macdonald, D., Seal, U. (eds.), 1998, Population and Habitat Viability Assessment for the African Wild Dog (*Lycaon pictus*) in Southern Africa. Final report from the Workshop, 13–17 October, 1997, Pretoria South Africa.

Naiman, R. J., Melillo, J. M., and Hobbie, J. E., 1986, Ecosystem alteration of boreal forest streams by Beaver (*Castor canadensis*), *Ecol.* **67:**1254–1269.

National Rivers Authority, 1995, River Habitat Survey: Field methodology guidance manual.

Nolet, B. A., 1994, Return of the Beaver to the Netherlands, Viability and prospects of a re-introduced population, PhD. Thesis, University of Groningen, Groningen.

Nolet, B. A., and Baveco, J. M., 1996, Development and viability of a translocated beaver (*Castor fiber*) population in the Netherlands, *Biol. Cons.* **75:**125–135.

Nolet, B. A., and Rosell, F., 1994, Territoriality and time budgets in Beavers during sequential settlement, *Can. J. Zool.* **72:**1227–1237.

Nolet, B. A., and Rosell, F., 1998, Comeback of the beaver *Castor fiber*: An overview of old and new conservation problems. *Biol. Cons.* **83**:165–173.

Ouderaa, A., van der., Fey, D., Stelt, J., van der, Zadelhoff, F.J., van der, 1985, Bevertochten, verslag van twee studiereizen in, 1984, Inspectie Natuurbehaoud Rapport Number, 1985 - 2, Staatsbosbeheer, Utrecht. [In Dutch, with English summary].

Perkins, S., 1998, An investigation into the ecological feasibility of reintroducing the European beaver (*Castor fiber*) into the Norfolk Broads. Unpub. report to The International Trust for Nature Conservation.

Rackham, O., 1986, *The history of the countryside,* J.M. Dent and Sons Ltd., London.

Romasov, B. V., 1992, Krankheiten der Biber, in: *Proceedings of the 2 International Symposium on Semiaquatic Mammals,* (Schröpfer, R., Stubber, M., and Heidecke, D., eds.), pp.130–140, Martin-Luther-Universität, Halle/Saale, Germany.

Rushton, S. P., Lurz, P. W. W., Fuller, R., and Garson, P. J., 1997, Modelling the distribution of the red and grey squirrel at the landscape scale: a combined GIS and population dynamics approach. *J. Appl. Ecol.* **34:**1137–1154.

Sage, B., 1981, The Beaver reintroduction project, *Discussion Papers in Conservation, University College London* **30:**46–49.

Sarrazin, F., and Barbault, R., 1996, Reintroduction - challenges and lessons for basic ecology, *TREE* **11:** 474–478.

SBB, 1983, Beavers in Nederland?, Staatsbosbeheer, Utrecht.

SNH, 1988, Reintroduction of the European beaver to Scotland. A public consultation. Scottish Natural Heritage.

Soule, M. E. (ed.), 1986, Conservation Biology: The science of scarcity and diversity. Sinauer, Sunderland, Mass.

South, A., Rushton, S., and Macdonald, D.W., in press, Predicting the outcome of a proposed reintroduction of the European Beaver (*Castor fiber*) to Scotland: a comparison of two modeling approaches, *Biol. Cons.*

Starfield, A. M., 1997, A pragmatic approach to modeling for wildlife management, *J. Wildl. Man.* **61**:261–269.

Stocker, G., 1985, The Beaver (*Castor fiber* L.) in Switzerland - Biological and ecological problems of re-establishment, *Swiss Federal Institute of Forestry Research Reports* **242**:1–149. [In German, with English summary].

Tattersall, F., and Macdonald, D. W., 1995, A review of the direct and indirect costs of reintroducing the European Beaver (*Castor fiber*) to Scotland, unpublished report to SNH, contract no. SNH/110/95/1BB, 28 April 1995.

Tyurnin, B. N., 1984, Factors determining numbers of the river Beaver (*Castor fiber*) in the European North, *Soviet J. Ecol.* **14(6)**:337–344.

UKCINC, 1979, Wildlife introductions to Great Britain: some policy implications for nature conservation, report of the United Kingdom Committee for International Nature Conservation; Working Group on Introductions.

Webb, A., French, D. D., and Flitsch, A. C. C., 1997, Identification and assessment of possible beaver sites in Scotland. Scottish Natural Heritage Research, Survey and Monitorin Report No. 94.

Yalden, D. W., 1986, Opportunities for reintroducing British mammals, *Mamm. Rev.* **16(2)**:53–63.

Zurowski, W., 1987, Differences in effects of the European Beaver reintroductions into lowland and mountainous tributaries of the Vistula river, *Transactions of the 18th Beaver congress*, Krakow.

Zurowski, W., 1988, Situation of the European Beaver in the Vistula river watershed and northeast Poland, *Investigations on Beavers* **7**:7–14.

Zurowski, W., and Kasperczyk, B., 1986, Characteristics of European Beaver population in the Suwalki lakeland, *Acta Ther.* **31**:311–325.

Zurowski, W., and Kasperczyk, B., 1988, Effects of reintroduction of European Beaver in the lowlands of the Vistula basin, *Acta Ther.* **33**:325–338.

13

BEAVER: A NEW PREY OF WOLVES IN LATVIA?

Comparison of Winter and Summer Diet of *Canis lupus* Linnaeus, 1758

Žanete Andersone

Ķemeri National Park, "Meža māja"
Ķemeri-Jūrmala, LV-2012, Latvia

ABSTRACT

The number of wolves in Latvia increased from 230 in 1985 to 997 in 1997. At the same time the population of wild ungulates declined significantly. The aim of this study was to examine the food habits of wolves in different seasons and to determine the ratio of wild ungulates in the diet. It is of special interest because no such study has ever been conducted in Latvia. Forty-one scats and 13 stomachs of hunted wolves were collected in the winter and summer of 1997. Scat analysis followed standard procedures. Microscopic analysis of hair was used to identify mammal species. Cervids (elk, red deer, roe deer) were not specified separately. Occurrence of particular food items (F%) and biomass of the prey consumed (B%) were calculated. In winter, the wolf diet consisted mostly of cervids (ca. 60%), wild boar, and livestock consumed as carrion. In summer, the ratio of cervids in scats was lower (48.5 F% and 31.4 B%) than it was in winter due to a more variable food content. Based on biomass beaver was the most important food item in summer (30.3 F% and 36.1 B%), followed by cervids, and wild boar (18.2 F% and 30.1 B%). Other food items (small mammals, birds, berries etc.) were of little importance. Thus, wild ungulates are the basic food for the Latvian wolf population in the both seasons. However, the sudden increase of beaver in the summer diet suggests switching by wolves to a more available prey during the depression of the ungulate populations.

1. INTRODUCTION

The wolf is the most common large carnivore that has always been present in Latvia since the last glaciation (Tauriņš, 1982). It is also the most negatively regarded predator

Beaver Protection, Managment, and Utilization in Europe and North America, edited by Busher and Dzięciołowski.
Kluwer Academic / Plenum Publishers, New York, 1999.

blamed for its depredation on livestock and wild ungulates (Kalniņš, 1943). A pilot study conducted in Latvia during the last several years found the main prey of wolves to be wild ungulates (cervids, especially roe deer), and wild boar (Andersone, 1998). Since the beginning of the 1990s, the wolf number in the country has increased significantly (Ozoliņš and Pilāts, 1995) until it reached a maximum of 997 individuals in 1997. In spring 1998, according to the official inventory, there were 908 wolves in Latvia. At the same time the ungulate population, which includes four species – elk, *Alces alces*, red deer, *Cervus elaphus*, roe deer, *Capreolus capreolus*, and wild boar, *Sus scrofa*, –declined significantly. During the same time period while the ungulate population has been declining, the beaver, *Castor fiber*, population has been increasing. The beaver is an autochthonous species that was eradicated in Latvia by over-hunting in the 19th century. The reintroduction of beaver in Latvia began in 1927 (Kalniņš, 1943; Balodis, 1990; Balodis, this volume), and the most current population estimate is approximately 50,000 individuals (Balodis, 1994). Balodis (this volume) estimates the beaver population in Latvia will be 100,000 individuals by the year 2000.

This study was designed to investigate the food habits of wolves in both winter and summer, and to document any seasonal difference in prey selection. I was especially interested in what role the expanding beaver population may play in the diet of wolves.

2. MATERIALS AND METHODS

Investigation of the wolf diet was carried out on a countrywide scale, and 41 wolf scats and 13 stomachs of killed wolves were collected in 1997 throughout Latvia. However, most of the scats and stomachs came from western and eastern Latvia, which are the regions with the highest wolf density.

Scat and stomach content analysis followed standard procedures (Goszczynski, 1974; Reynolds and Aebischer, 1991). Identification was based on microscopic analysis of hair structure (Day, 1966; Teerink, 1991). Cervid species were not separated during the analysis due to their very similar hair structure. Scats were put into water with detergent for 24 hours, then rinsed through fine mesh (0.5 - 1 mm) to separate undigested remains (bones, skin fragments, hair etc.) which were later dried at + 40 C. Dried samples were weighed (1 g precision) and multiplied with Floyd's (1978) coefficients to determine the amount of the biomass consumed. The percent biomass was calculated by the following equation where B_i is the biomass of species i and B_t is the total biomass:

$$B\% = \frac{B_i}{B_t} \bullet 100$$

For the stomach content analysis, biomass of a prey consumed was weighed directly.

Frequency of occurrence was calculated as a percentage from the total number of scats. In the following equation F_i is the frequency of species i, and F_t is the total frequency:

$$F\% = \frac{F_i}{F_t} \bullet 100$$

Due to a limited number of samples the results from scat and stomach analysis were combined.

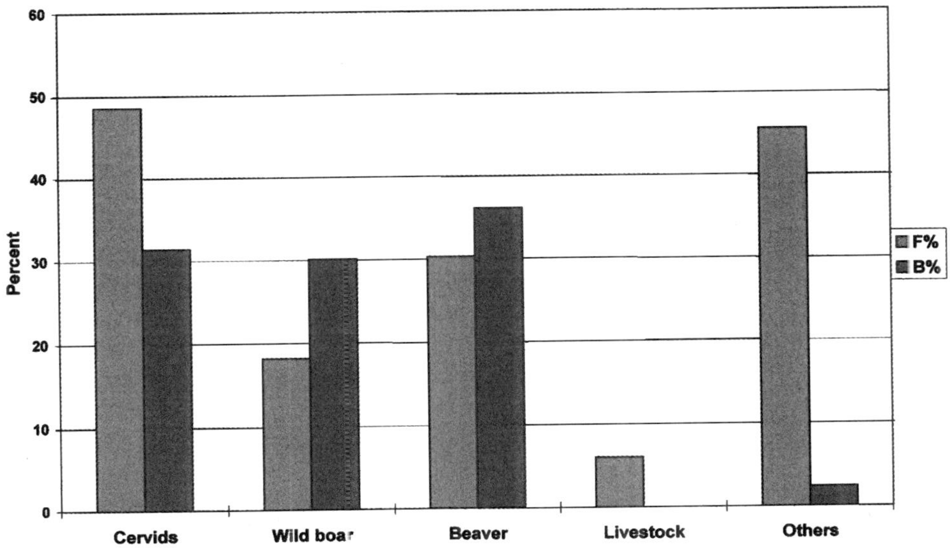

Figure 1. Summer diet of the wolf *Canis lupus* in Latvia (n=33). F%—frequency of occurence; B%—percentage of biomass consumed.

3. RESULTS

The current analysis confirmed that wild ungulates are the basic food for wolves in Latvia in both summer and winter seasons (Figure 1, Figure 2). However, it also has revealed the importance of a new prey—the beaver (Figure 1).

In this study, beavers were documented in the wolf diet for the first time. Beaver remains were found in wolf scats collected in western Latvia (n = 8) and in stomachs of a wolf pack (female with three pups) harvested in the eastern part of the country (n = 4). In June 1998, in eastern Latvia, a lactating female wolf was shot, and a young beaver (one to two years old) was found in her stomach. This female was shot while attacking sheep in a cot at night. Obviously, a beaver was not enough food for the litter, although this example illustrates that a pack with yearlings can successfully feed on beavers.

Beavers were only a part of the wolf diet in spring and summer. Beavers constituted a considerable part of the wolves' summer diet (30.3% frequency of occurrence, 36.1% of biomass consumed) (Figure 1). Most of the scats containing beaver remains were found in the Slītere Nature Reserve where no wolf predation on beavers had been recorded in previous years (Andersone, 1998). This difference in wolf predation on beavers between previous years and the current study was most likely caused by the time lag due to the wolves' adaptation to a new prey.

Beavers were not found in winter diet of wolves (Figure 2), although this may be due to a small sample size (n = 21). It is also possible that conditions of thick ice and snow cover can successfully protect beavers from wolf predation A significant correlation was found between the population dynamics of wolves and beavers during the last 30 years (Figure 3) which possibly indicates strong mutual relations between the two species.

4. DISCUSSION

Other authors, both in North America and in Eurasia, have reported that wolves prey on beavers when the latter are available (Ballard *et al.*, 1987; Bibikov, 1985). However,

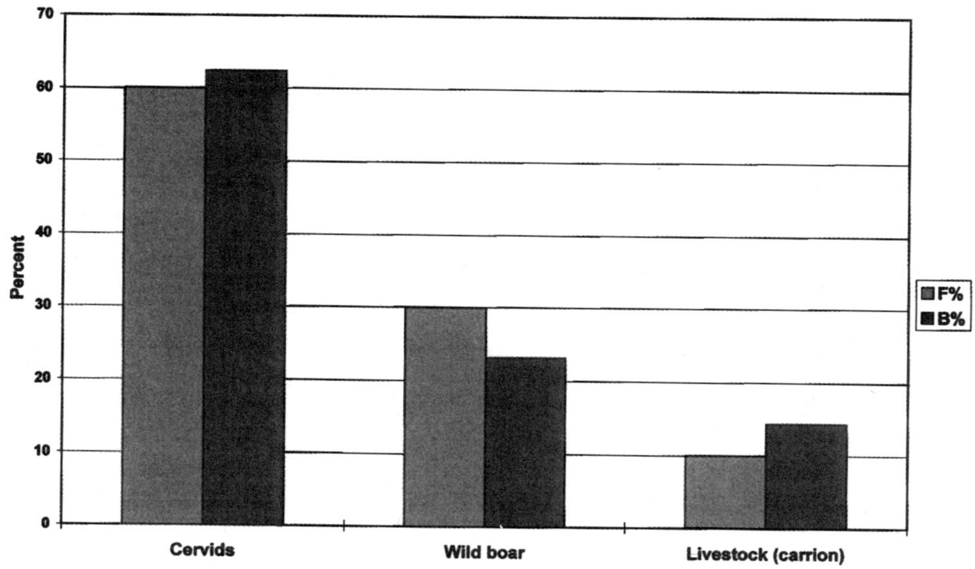

Figure 2. Winter diet of the wolf *Canis lupus* in Latvia (n=21). F%—frequency of occurence; B%—percentage of the biomass consumed.

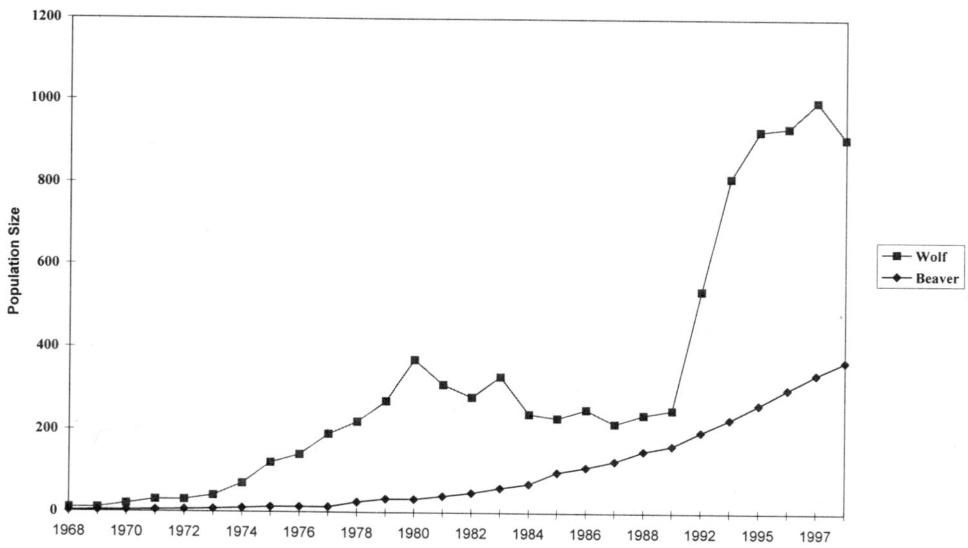

Figure 3. Beaver and wolf population dynamics, 1968–1998. The number of beavers in given in hundreds.

documented wolf predation on beavers has not been reported for Latvia, which might be due to the lack of previous studies on wolves. However, it is highly probable that these results indicate a real change of prey-predator relationships in nature.

In Europe, wolves are reported to prefer wild ungulates—cervids and wild boar—as the main prey (Jedrzejewski *et al.*, 1992; Okarma, 1995), and this is also the case in Latvia (Andersone, 1998; Andersone, this study). It appears that the significant decrease in ungulate numbers in the 1990s induced wolves to switch to an alternative and more available prey (beavers), at least during the summer.

Because the collapse of the fur market made beaver trapping unprofitable, the wolf is practically the only factor limiting the growth of the beaver population. High beaver numbers can cause damage to agriculture and forestry, and control of the beaver population is necessary. Historically, wolves were one of the important natural factors controlling the number of beavers in the wild, and continued wolf predation on beavers is a desirable management objective. On the other hand, the wolf is considered a pest species that is hunted without any limitations. The official policy is to reduce the number of wolves as much as possible, and more than 300 wolves were harvested annually during the last three hunting seasons. The number of wolves would naturally decline due to food shortage (low density of cervids and wild boar). However, if wolves use beavers as an alternative prey in the summer, this could delay or alter the wolf population decline. In winter, the energetic cost of hunting beavers is probably too high, and wolves must rely on wild ungulates as the main prey during this season. Thus, beavers, being a supplementary food item for wolves in the summer, can support survival rates of wolf pups at a stable level even during the current ungulate depression. This in turn can affect ungulate populations by keeping them at low densities for a longer time than if such an alternative prey was not available. However, to test this hypothesis further research and long-term monitoring of the predator-prey relationships are needed.

REFERENCES

Andersone, Ž., 1998, Summer Nutrition of the Wolf (*Canis lupus*) in the Nature Reserve Slītere, Latvia, *Proc. Latvian Acad. Sci., Section B,* **52:**1–2.

Balodis, M, 1990, [The Beaver. Biology and Management in Latvia], Zinātne, Rīga, p. 271. [In Russian with English summary]

Balodis, M., 1994, Beaver Population of Latvia: History, Development and Management, *Proc. Latvian Acad. Sci.. Section B,* **7/8:** 122–127.

Ballard, W. B., Whitman, J. S., and Gardner, C. L., 1987, Ecology of an exploited wolf population in south-central Alaska, *Wildlife Monographs*, **98:** p. 54.

Bibikov, D. I. (ed.), 1985, [*The Wolf. History, Systematics, Morphology, Ecology*], Izd. Nauka, Moskva: p. 606. [In Russian]

Day, M. G., 1966, Identification of hair and feather remains in the gut and faeces of stoats and weasels, *J. Zool.*, **148:** 201–217.

Floyd, T. J., Mech, L. D., and Jordan, P.A., 1978, Relating wolf scat content to prey consumed, *J. of Wildl. Manage.*, **42 (3):** 528–532.

Goszczynski, J. 1974. Studies on the Food of Foxes. Acta theriol., 19 (1): 1–18.

Jedrzejewski, W., Jedrzejewska, B., Okarma, H., and Ruprecht, A. L., 1992, Wolf predation and snow cover as mortality factors in the ungulate community of the Bialowieza National Park, Poland, *Oecologia*, **90:** 27–36.

Kalniņš, A., 1943, [*Huntsmanship*], Latvju grāmata, Rīga, p. 704. [In Latvian] Okarma, H., 1995, The trophic ecology of wolves and their predatory role in ungulate communities of forest ecosystems in Europe, *Acta theriol.*, **40 (4):** 335–386.

Ozoliņš, J., and Pilāts, V., 1995, Distribution and status of small and medium-sized carnivores in Latvia, *Ann Zool. Fennici*, **32:** 21–29.

Reynolds, J. C., and Aebischer, N. J., 1991, Comparison and quantification of carnivore diet by faecal analysis: a critique, with recommendations, based on a study of the Fox *Vulpes vulpes*, *Mammal. Rev.,* **21 (3):** 97–122.
Tauriņš, E., 1982, [*Mammals of Latvia*], Zvaigzne, Rīga, p. 256. [In Latvian]
Teerink, B. J., 1991, *Hairs of West European Mammals*, Cambridge University Press, Great Britain.

14

UNDERSTANDING NORTH AMERICAN BEAVER BEHAVIOR AS AN AID TO MANAGEMENT

Bruce A. Schulte[1] and Dietland Müller-Schwarze[2]

[1]Department of Biology
Providence College, Providence, Rhode Island 02918
[2]Department of Biology
SUNY-CESF, Syracuse, New York 13210

ABSTRACT

In the twentieth century, the North American beaver, *Castor canadensis*, has rebounded from widespread extirpation to high abundance. Swift increases in number and range expansion, coupled with rapid human population growth and development, has resulted in beaver being labeled a nuisance species in many areas. Major problems created by beavers include cutting valuable trees, altering water flow primarily through building dams, and subsequent flooding of properties and roads. Control efforts have focused on harvesting beavers, fencing of trees, removal of dams and water flow devices in dams; less explored efforts comprise beaver exclusion and fertility control. Beavers provide many benefits, including the creation of wetlands, raising water tables and providing human recreational and aesthetic pleasures. Ecologically, beavers are considered keystone species and ecosystem engineers. Beavers are primarily nocturnal, monogamous, relatively long-lived rodents that reside in family groups. Beaver activity varies seasonally and by age class. Evidence exists for different roles or different timing of roles played by the sexes and by age class. Individual behaviors and population responses to changing conditions appear highly plastic. Timing of dispersal and site relocations can span six months or more. Beavers may settle in streams, lakes or reservoirs. Individuals appear to learn rapidly to avoid or not avoid particular objects (e.g., live-traps). However, responses to beaver scent are largely non-habituating, yet can be variable depending on the type of scent and the age/sex of the responder. Because of this highly malleable behavioral repertoire, beavers may be managed best by the alteration of human behavior and attitudes.

Beaver Protection, Managment, and Utilization in Europe and North America, edited by Busher and Dzięciołowski.
Kluwer Academic / Plenum Publishers, New York, 1999.

1. INTRODUCTION

1.1. History of North American Beaver Population

The North American beaver, *Castor canadensis*, played an important role in the settlement of North America (Allred, 1986; Hilfiker, 1990). The beaver pelt was a valuable commodity for hundreds of years and stimulated the westward exploration of North America by Europeans. During the early 1600s, beavers occupied the vast majority of waterways in North America (Jenkins and Busher, 1979; Ruedemann and Schoonmaker, 1938), and beaver numbers then have been estimated at 60–400 million (Naiman, Melillo and Hobbie, 1986; Naiman, Johnston and Kelley, 1988). By 1850, beaver numbers were greatly diminished (Jenkins and Busher, 1979). Fortunately for beavers, fashions changed in Europe in the mid-1800s and silk replaced beaver felt as the vogue material for hats (Allred, 1986). Trapping of beavers remained heavy in some regions, such as parts of Canada, but in most of North America the late 1800s and early 1900s was a period of slow recovery for beaver populations. Remnant populations, such as the 5–15 beavers that survived in the Adirondack Mountains of New York State (Ermer, 1988; Johnson, 1927; Radford, 1907), were supplemented by reintroductions from western populations (Larson and Gunson, 1983). The population of North American beavers was estimated at 6–12 million individuals by the 1980s (Naiman *et al.*, 1986, 1988).

This extreme fluctuation in beaver numbers over several centuries has been accompanied by a rapid increase in sympatric human numbers (Table 1). In 1600 the human population in the United States and Canada was approximately one million; by 1850 this number had risen to 26.5 million, and today there are nearly 300 million people in the United States and Canada (McEvdy and Jones, 1978; Roberts, 1998). Hence, the application of our understanding of beaver behavioral ecology to the management of beavers must be done in light of historical beaver numbers and the current human population. In this paper, we describe some components of the population biology of beaver, followed by benefits and problems created by beaver. These topics provide a background for examining the behavioral ecology of beavers in terms of their ecological roles and their behavioral repertoire. We conclude with a discussion on how this knowledge can aid in the management of beavers.

1.2. Range of the North American Beaver Population

The biological idea of range has two primary components, geographic and ecological (Ricklefs, 1997). In the 1600s, beavers were distributed throughout North America ex-

Table 1. Population estimates (in millions) for beavers and humans in the United States and Canada since 1600

Date	Beaver	Human
1600	60-400[a]	1[b]
1850	1-5?	26.5[b]
1900	<<1?	81[b]
Present	6-12[a]	300[c]

[a]Naiman *et al.* 1986.
[b]McEvdy & Jones 1978.
[c]Roberts, 1998.

cept for southwest regions of desert, the peninsula of Florida and the tundra areas of Alaska and Canada (Hill, 1982; Jenkins and Busher, 1979). Following the precipitous decline of their population due to trapping by humans, beavers were found mostly in remote regions of western North America. East of the Mississippi River beavers were extirpated or survived in only small remnant populations (Johnson, 1927; Johnson and Chance, 1974; Larson and Gunson, 1983). Today, beavers have become reestablished in much of their former geographic range. For example, in recent years beavers have expanded their range from northern to southern Louisiana (Elsey and Kinler, 1996) and have been found using caves in northern Florida (Gore and Baker, 1989).

In terms of ecological range, beavers live in riparian habitats. Beavers are limited in their distribution by their access to water and in some cases their ability to impound water (Novak, 1987). In assessing appropriate habitat features, physical properties have been considered of primary importance (Allen, 1982; Howard and Larson, 1985; McComb, Sedell and Buchholz, 1990; Slough and Sadleir, 1977), especially when dam building is necessary. For instance, beavers are unlikely to settle in streams with gradients exceeding 15% (Allen, 1982; Howard and Larson, 1985), and cross-sectional area is an important determinant of dam location (Beier and Barrett, 1987; McComb *et al.*, 1990). Broad, deep streams provide for year-round maneuverability and food storage, while narrower streams are easier to dam (Barnes and Mallik, 1997). Most studies in the past 40 years have agreed that shoreline food availability is not a critical criterion for determining beaver settlement. If beavers can dam streams, then they can reach suitable food resources nearby. However, vegetative characteristics affect where beavers will construct lodges, which are generally built in areas with greater cover and small to medium sized trees (Dieter and McCabe, 1989). Furthermore, recent studies by Barnes and Mallik (1996, 1997) indicate that streamside vegetation is an important habitat factor influencing beaver dam establishment. Specifically, the species composition, stem size and spatial distribution of woody vegetation along streams were strong determinants of dam location in northern Ontario watersheds (Barnes and Mallik, 1997). Beavers are not impelled to create dams and can live for many years in larger rivers or lakes (Buech, 1995; Smith and Peterson, 1991; Wheatley, 1997c). Although human activities have altered drastically the waterways and wetlands in North America, beavers are reclaiming much of their former dominion over the year-round lotic and lentic regions of North America.

1.3. Beaver Life History Characteristics

The basic life history characteristics of beavers have been summarized in detail elsewhere (e.g., Allred, 1986; Bradt, 1938; Hilfiker, 1990; Jenkins and Busher, 1979). However, for our purposes some aspects need to be restated. Beavers are monogamous, although pair bonds may last only a few years because of mortality of one individual (Svendsen, 1989). Family units are composed of an adult pair and progeny from two successive years, averaging 3–8 beavers per site (Bergerud and Miller, 1977; Hodgdon, 1978; Novak, 1977; Schulte, 1993). Such family units have traditionally been called a "colony" (Bradt, 1938), although there may be much variation to a 'typical' colony unit (Townsend, 1953). As population density increases, the composition of beaver colonies is altered. Two-year olds often delay dispersal, resulting in more than two adult beavers per colony, and beavers may move more among colonies (Bergerud and Miller, 1977; Busher, 1987; Busher, Warner and Jenkins, 1983; Müller-Schwarze and Schulte, this volume; Swenson, Knapp, Martin and Hinz, 1983; Van Deelen and Pletscher, 1996).

Mating takes place in the winter months and births generally occur from April through June (Jenkins and Busher, 1979), although parturition may occur as early as February and as late as November (Bradt, 1939; Wigley, Roberts and Arner, 1983). Young-of-the-year, called kits, possess a mixture of altricial and precocial characteristics (Patenaude, 1983; Patenaude and Bovet, 1983). Typically, beavers disperse in the spring of their second year when they establish a new active site and mate the following winter (Brenner, 1964; Leege, 1968; Svendsen, 1980a,b). Dispersal distance varies considerably but is commonly 8–10 km (Leege, 1968; Sun, Müller-Schwarze and Schulte, unpublished; Van Deelen and Pletscher, 1996). However, movements of several hundred kilometers have been recorded (Hodgdon, 1978; Libby, 1957). Beavers are long-lived for rodents with recorded ages of 24 years in the wild (Brown, 1979) and twice that long in captivity (Nowak, 1991). Fertility is highest for beavers 2.5–7 years of age with 2–8 kits (average 3–4) born each year, but yearling beavers and older beavers can reproduce (Henry and Bookhout, 1969; Leege and Williams, 1967; Lyons, 1979; Payne, 1984b). Sexual maturity may be suppressed in the presence of dominant adults (Brooks, Fleming and Kennelly, 1980; Novak, 1977, 1987; Svendsen, 1980a; Wilsson, 1971). However, the validity of this hypothesis is untested, and the mechanism for its action has not been adequately investigated.

1.4. Beaver Population Dynamics

Because of the variability in reproductive output, beavers have the potential to rapidly take advantage of high quality habitats, increasing population numbers in a relatively short time-span (Boyce, 1981; Larson and Gunson, 1983; Payne, 1984a,b). Although general principles of furbearer population management, including beaver, are well-described (Novak, 1987; Wolfe and Chapman, 1987), few attempts have been made to model beaver population dynamics (Fryxell, unpublished; Molini, Lancia, Bishir, and Hodgdon, 1981). Yearling beavers are often abundant in a population and can make a significant contribution to the overall reproductive output of a population in a given year (Payne, 1984b). In heavily trapped populations, beaver "colonies" are composed of more single individuals and pairs of beavers, while in untrapped populations there are more typical family units (Leege and Williams, 1967; Payne, 1982; Peterson and Payne, 1986). Under such conditions, beavers are likely to reproduce earlier and at a smaller size, sacrificing longevity for early reproductive success (Boyce, 1981). Reproduction may occur in the natal site because adults have suffered mortality. In addition, yearlings may disperse and reproduce (Jenkins and Busher, 1979). Some beavers may live for a time as transients or floaters (Allred, 1981; Van Deelen and Pletscher, 1996), and this number is likely to increase with population size. Molini *et al.* (1981) estimated that as much as 20% of a population might be floaters. Transients may serve as a ready source of breeders following mortality events. The fate of this transient population at times of very high density (i.e., at carrying capacity) is uncertain. Potential transients may remain at or return to natal sites to help and thereby increase their inclusive fitness. In addition, transients may join an already established adult pair, possibly to inherit part or all of the territory in the future. Finally, transients may occupy suboptimal habitats and await opportunities to breed with or without permanently occupying an established territory either due to loss of a same-sexed adult, or copulation with a mated individual, respectively.

A critical issue in the management of beavers has been determining the relationship between current beaver numbers and potential beaver occupancy of a region. Typically, aerial surveys have been used to identify beaver impoundments (Brown and Parsons,

1982; Dickinson, 1971; Howard and Larson, 1985; Parsons and Brown, 1978). Vegetative analysis from wetland inventory maps and land capability classification systems also have permitted comparable evaluation of potential beaver habitat to those achieved from aerial photography (Gotie and Jenks, 1984; Slough and Sadleir, 1977). However, habitat evaluation alone often cannot distinguish between active and inactive sites. To determine population levels, food caches are counted in the fall (Bergerud and Miller, 1977; Hay, 1958; Payne, 1981; see also Busher and Lyons this volume). Only active colonies (but not all) have food caches and population size can be estimated by using an average number of individuals per colony multiplied by the number of caches. This method can be inaccurate where colony size changes abruptly from year to year (Swenson *et al.*, 1983). Estimating colony size for a particular population can be extremely difficult and perhaps more so with European beavers because they seem to be harder to livetrap (Parker, personal communication). The area of the food cache was a useful index for colony size in a Montana study (Easter-Pilcher, 1990). In southerly climates where food caches are not created, counting caches is not applicable, and other assessments of habitat are necessary to estimate beaver numbers. More complex assessments of habitat have been used to estimate beaver numbers with varying success. Use of a habitat suitability index did not provide reliable information on beaver population size in eastern Kansas (Robel, Fox and Kemp, 1993). However, using multiple stepwise regression, Broschart Johnston and Naiman (1989) were able to predict beaver density based upon characteristics of beaver impoundments in the Voyageurs National Park, Minnesota. This research revealed that the rate of establishment of beaver ponds was higher during the first 20 years of population growth than the next 20 years (Johnston and Naiman, 1990a,b). Hence, the impact of beavers on the landscape changes with population growth rate and size.

2. MANAGING BEAVERS: THE GOOD AND THE BAD

2.1. Benefits of Beavers

Beavers alter the landscape through their construction of dams, carving of canals, cutting of trees and burrowing into banks (Naiman *et al.*, 1986, 1988). Beavers expand riparian habitat. Their actions lead to an increase in aquatic structural diversity and overall production (Naiman *et al.*, 1986; Wilde, Youngberg and Hovind, 1950). Such alterations are beneficial to waterfowl and other waterbirds for breeding and feeding grounds (Beard, 1953; Brown, Hubert and Anderson, 1996; Brown and Parsons, 1979; Novak, Baker, Obbard and Mallock, 1987; Renouf, 1972). Active beaver ponds tend to have a higher diversity of birds than inactive sites (Medin and Clary, 1990), primarily because of the diversity of habitats created and maintained by beavers (Brown *et al.*, 1996; Grover and Baldassarre, 1995; Ringelman, 1991). In addition, the benefit to waterfowl of beaver presence increases with time. In a study comparing regions with beavers that were trapped or untrapped, water surface area and total wetland habitat increased after a 2–3 year cessation of trapping (McCall, Hodgman, Diefenbach and Owen, 1996). Waterfowl use of wetlands and density of waterfowl (e.g., black ducks, Canada geese and mallards) followed this expansion of habitat, but with a lag of one to several years (McCall *et al.*, 1996). Yet, despite the aesthetic and potential economic gains derived from thriving waterfowl and waterbird populations, such singular values are rarely enough to override the potential costs to humans of nearby beaver occupation (Müller-Schwarze, 1996). However, the presence of beavers offers a much wider range of benefits, befitting of even an economic viewpoint of nature (Costanza *et al.*, 1997). We list additional benefits of beavers and their actions below (Naiman *et al.*, 1986; Olson and Hubert, 1994; Woodward, 1983):

1. Beaver activity increases the resistance of ecosystems to disturbance.
2. Beaver dams serve to stabilize the water table and water flow throughout the year.
3. Beaver ponds serve as water sources for irrigation, livestock watering and fire fighting.
4. Sediments are retained, resulting in enhanced water quality and aiding erosion control.
5. Organic materials are retained, increasing aquatic productivity that enhances fish growth.
6. Cutting of trees by beavers opens up the canopy, resulting in growth of herbaceous forage and resetting succession.
7. Beaver sites are a source of environmental education and of recreation such as hunting, fishing, birding and photography.

2.2. Problems with Beavers and Control Methods

The major alteration in landscape exacted by beavers also is the reason beavers create problems for humans (Payne and Peterson, 1986). The construction of dams alters the flow rate and course that water travels. The flooding of previously dry areas may cause problems with associated economic costs (Ermer, 1988). In the later half of the twentieth century human and beaver populations have risen dramatically (Table 1). As humans expand their dwellings and roadways into previously rural regions, interactions with beavers will continue to increase.

Some general problems associated with beavers and their actions are listed below (Müller-Schwarze, 1996; Naiman *et al.*, 1986; Olson and Hubert, 1994; Woodward, 1983):

1. Beaver dams cause flooding, eliminating habitat and damaging resources used by humans such as timber, crops and roads.
2. Beavers cut and girdle trees, killing ornamentals and income-based timber.
3. Beaver dams may adversely affect some aquatic species by altering water depth, flow and temperature (McDowell and Naiman, 1986).
4. Beaver ponds may harbor infectious organisms such as *Giardia lamblia* (Dixon, Parenteau, Marinteau and Fournier, 1997) and may provide breeding areas for mosquitoes (Hudson, 1978), but beavers also may decrease habitat for mosquito larval development (Butts, 1988).
5. Beaver ponds may be a source of methane, possibly adding to global warming (Ford and Naiman, 1988).

Control methods for problems caused by beavers can be categorized into two types: population level and site-specific. In population level control, beaver numbers are maintained at a level deemed suitable for the habitat and human population of the region (e.g., 0–35% of potential occupancy, Ermer, 1988). The hope is that beavers will reside mostly in areas that do not conflict with human uses. However, even when population control is enacted, beavers may become a nuisance in a particular site. The second level of control is site specific, whereby attempts are made to either keep beaver out of an area or to minimize damage by beavers at a site. In many instances, beavers are welcomed at a site but certain activities (e.g., cutting of particular trees or flooding of some areas) are unwelcome. In such cases, very specific measures can be taken to alleviate or prevent problems.

Below, we list the major forms of control that either have been or could be used to reduce damage from beavers at each of these levels (Müller-Schwarze, 1996; Olson and Hubert, 1994; Woodward, 1983). We discuss some of these further in section 4.0 on management of beavers.

A. Population Control

1. Beavers are harvested (by trapping or shooting) to maintain lowered numbers.
2. Beavers are livetrapped and relocated to a distant location.
3. Beavers are sterilized or fitted with a fertility control hormone implant.
4. The population level of natural predators is increased (Arner, Mason and Perkins, 1981; Smith, Trauba, Anderson, and Peterson, 1994; Thurber and Peterson, 1993).
5. The application of repellents (predator odors, territorial odors) is used to prevent dispersing beaver from settling in a particular region.

B. Site Control

1. Individual colonies of so-called nuisance beavers are harvested completely.
2. Individual colonies of beavers are harvested at sustainable levels.
3. Nuisance beavers are livetrapped and relocated to a nearby region.
4. Adult beavers are castrated or fitted with fertility control implants and maintained at the current site.
5. Resources are physically protected to prevent damage. For instance, trees can be encircled with fencing, bank areas can be fenced to reduce accessibility and culverts can be screened to avoid plugging.
6. Repellents are used to prevent the cutting of particular trees (antifeedants) or to keep beavers from visiting particular areas (e.g., dogs may stand guard in sensitive places).
7. The installation of water control devices in dams can maintain adequate water levels for beavers while avoiding flooding of valuable resources (Buech, 1985; Roblee, 1984).

3. BEHAVIORAL ECOLOGY OF BEAVERS

3.1. Ecological Roles

Beavers serve two major ecological roles: as ecosystem engineers (Jones, Lawton and Shachak, 1994, 1997), and as keystone species (Naiman *et al.*, 1986; Paine, 1966, 1969), in their riparian ecosystems. Ecosystem engineers are "organisms that directly or indirectly control the availability of resources to other organisms by causing physical state changes in biotic or abiotic materials. Physical ecosystem engineering by organisms is the physical modification, maintenance or creation of habitats" (p. 1947, Jones *et al.*, 1997) The activities of ecological engineers, like beavers, affect other species by altering resource availability either directly or indirectly (Jones *et al.*, 1997; Naiman *et al.*, 1988; Naiman, Pinay, Johnston and Pastor, 1994).

Beavers add organic material to the aquatic environment (Naiman and Melillo, 1984; Naiman *et al.*, 1986) through their actions as central-place foragers (Basey, Jenkins and Busher, 1988; Fryxell and Doucet, 1991). Beaver dams aid in the retention of this matter and a beaver-modified section of stream has a thousand fold increase in nitrogen relative to unaltered stream (Naiman and Melillo, 1984). Beavers have a dramatic impact on the carbon cycle in riparian habitats (Naiman *et al.*, 1986) and the evasion rate of methane is much higher in beaver ponds than other free flowing sections of stream (Ford and Naiman, 1988). The woody debris and sediment entrapped at beaver sites, including near lodges at sites on boreal lakes, has a significant impact on the littoral fauna as well (France, 1997). Beaver activity alters stream fish assemblages and species populations, generally in an ecologically positive direction (Snodgrass and Meffe, 1998). In addition, beaver sites along streams alter chemical aspects of the water and soil (e.g., pH, fertility) and biotic aspects of a stream (Naiman *et al.*, 1994; Schlosser, 1995; Smith, Driscoll, Wyskowski, Brooks and Cosentini, 1991; Wilde *et al.*, 1950).

The foraging of beavers can have a dramatic impact on the trophic dynamics in a community. The term keystone species, although often broadly applied, generally is limited to species that have a significant impact on community composition through trophic interactions. Food selection by beavers has short-term and long-term impacts on the species composition of a forest (Johnston and Naiman, 1990c). Beavers remove entire trees from the forest and can virtually clearcut regions of forest near their sites (Johnston and Naiman, 1987). This creates patches in the forest and allows understory vegetation to grow (Olson and Hubert, 1994). Johnston and Naiman (1990c) state that beavers remove a greater proportion of biomass per foraging area than any other herbivore. The preferences of beavers for particular tree species has ramifications for rates of forest succession and future composition as well (Barnes and Dibble, 1988; Johnston and Naiman, 1990c; Neff, 1957; Wilde *et al.*, 1950). The congener, *C. fiber*, has similar impacts on riparian habitats in Europe (Nolet, Hoekstra and Ottenheim, 1994).

Beavers also exert indirect effects in their trophic webs. Regrowth of cottonwood trees (*Populus* spp.) following cutting by beavers results in trees with much higher levels of phenolic glycosides (Martinsen, Driebe and Whitham, 1998). These chemicals serve as a defense against mammalian herbivores, but leaf beetles (*Chrysomela confluens*) are attracted to the regrowth. Leaf beetles sequester the chemicals for their own defense and develop faster on a diet of resprouted cottonwood compared to first growth. Hence, from herbivore-herbivore indirect interactions to landscape level alterations, beavers are ecosystem engineers and keystone species in a habitat that is impoverished in much of North America.

3.2. Beaver Behavior

The behavior of beavers has been studied extensively, although primarily in more northerly latitudes of their range (e.g., Busher and Jenkins, 1985; Hodgdon and Lancia, 1983; Schulte, 1998; Sun, 1996; Tevis, 1950; Wheatley, 1997a-c). Basey (this volume) discusses foraging in depth, so we will not consider it here. We also will not cover topics related to the bioenergetics of beavers (e.g., Bovet, 1994; MacArthur and Dyck, 1990, 1994; Smith and Jenkins, 1997). Our emphasis will be on behavioral characteristics of beavers that might impact their management by humans.

The periods of activity and the roles played by different ages and sex of beaver can affect actions taken in population or site management of beavers. For example, when attempting to control population size by harvesting beavers or by fitting beavers with a fertility control implant, knowing activity patterns and capturing females is very important. In general, beavers are active between 17:00–08:00 hr (Buech, 1995; Busher, 1975, 1980; Hodgdon and Lancia, 1983; Lancia, 1979; Tevis, 1950). Most of this time is during dark-ness, but in very northerly latitudes some daylight may prevail for much of this active period during the summer (Wheatley, 1997a). Home range size varies by season and latitude interactively. For instance, beavers in Manitoba had the largest home range during the summer (Wheatley, 1997c), whereas beavers in South Carolina utilized the smallest home range in summer (Davis, 1984). Adult females had smaller home ranges than the rest of the family during summer only (Wheatley, 1997b). In a study by Buech (1995), adult beavers living in Minnesota lakes spent most of their activity period feeding, traveling and in the lodge. Adult females spent more time feeding than males in late spring and summer, while males traveled more. Overall, females functioned primarily as caregivers to the young, whereas males served in roles of protection and provisioning of kits, construction and maintenance (Buech, 1995). Similar results were found for beavers in the Sierra Nevada range of California where adult males spent twice as much time as adult females in maintenance and food-caching activities (Busher, 1980; Busher and Jenkins, 1985). Clearly, only the adult female can provide milk to kits following birth in the late spring or early summer, so the adult female must be viewed as the primary caregiver. Hodgdon and Larson (1973) found that adult females performed more tasks of maintenance and food-cache building, but males inspected dams more often. They described the adult female as the "alpha member" of the beaver family in their Massachusetts' study. Brady and Svendsen (1981) stated that adult males patrolled the site, added to the food cache and provisioned kits more often than females. However, they noticed no dominance hierarchy within the family that they observed in Athens County, Ohio. Busher (1980, 1983) did not observe a gender hierarchy in his California population, but did find older animals to be dominant to younger animals. Overall, it appears that adult males and females differ somewhat in their activities and daily range of travel, but these differences are reduced in the fall compared to the summer (Buech, 1995; Busher and Jenkins, 1985; Wheatley, 1997b).

Spring and early summer is the main period of dispersal by beavers, especially two-year-olds (Svendsen, 1980a). Therefore, the activity pattern of juveniles (1- and 2-year olds) compared to adults also is of interest. The above studies and others generally concur that adult beavers spend more time in maintenance, patrolling and provisioning of kits than do juvenile beavers. Juveniles spend time feeding, swimming and in social interactions. The behavior of yearling females has been likened more to that of adult females, while yearling males behaved more like kits, suggesting females may mature faster (Busher and Jenkins, 1985). The success of livetrapping beavers is one indicator of differences in adult versus juvenile activity patterns. Rosell and Kvinlaug (1998) have reviewed studies reporting on the success of livetrappping beavers. In general, beavers are easier to catch in Hancock live-traps during the first few days of trapping than on subsequent days. Juvenile beavers are captured more easily than adult beavers and fewer trap nights are required to catch beavers the first time than beavers that have been trapped previously. Despite the supposition that beavers are not highly reliant on vision (Doucet, Walton and Fryxell, 1994; Hodgdon and Lancia, 1983), success at trapping is higher with than without camouflaged traps. Vegetation used as camouflage may cover odor cues as well as visual ones.

Our own studies at Allegany State Park in western New York State do not support some of these generalizations. Our trap data shows that beavers from four different age classes were recaptured in proportion to their abundance in the population (Table 2, P^2=

Table 2. Number of beavers by age class and sex livetrapped at Allegany State Park in 1985, 1987, 1989–92 and recaptures within the same age class (during same trapping season or same year such as spring to fall)

	Total captured			Recaptures—same age (same season or year)			
Age class	Male	Female	Total	Male	Female	Total	Expected[a]
Adult	36	34	70	4	5	9	11.98
Two	24	19	43	3	1	4	7.36
One	42	36	78	12	7	19	13.35
Kit	20	11	31	2	4	6	5.31
Totals	122	100	222	21	17	38	38

[a]Expected based on proportion of total captured for age class multiplied by total number of recaptures (e.g., (70/222)*38 = 11.98). Comparison of total to expected recaptures was nonsignificant (P^2= 4.76, *df* = 3, *P* = 0.19).

4.76, *df* = 3, *P* = 0.19). In addition, there were no differences in the recapture rate of males and females overall (Table 3A, P^2= 1.32, *df* = 1, *P* = 0.25) or for adult males and females that were both captured and recaptured as adults (Table 3B, P^2= 1.33, *df* = 1, *P* = 0.25). However, the latter instance could be the result of a rather small sample size. Adult males were recaptured twice as often as adult females despite the nearly identical numbers in the population (36M:34F). A higher recapture rate for males could be indicative of the greater role that male beavers play in patrolling and defending their colony site.

When trying to catch beavers, traps are set in areas that show obvious signs of beaver activity such as a lodge and well-maintained dams. Hence, beavers that are dispersing or transient are rarely targeted for capture. Wheatley (1997b) found that non-reproductive beavers had larger home ranges than beavers with kits. Trapping non-reproductive beavers may be more difficult because a larger area must be covered. However, for population and site control, capturing or displacing non-reproductive beavers may be easier and more effective than evicting an established, territorial beaver family.

Beavers of both species mark and defend their sites (Aleksiuk, 1968; Bergerud and Miller 1977; Müller-Schwarze and Heckman 1980; Rosell and Nolet, 1997; Rosell, Bergan and Parker, 1998; Schulte 1998; Sun 1996; Svendsen 1980a; Wilsson, 1971). Aggression between family members is thought to be rare, but interactions between individuals

Table 3. Number of (A) male and female beavers livetrapped at Allegany State Park in1985, 1987, 1989–92 and those recaptured, and (B) number of adult beavers recaptured following a previous capture as adults

Beavers	Female	Male
A. All age classes of beavers		
Total captured	100	122
Recaptured[a]	28	49
Expected recaptures	35	42
B. Adult beavers only		
Total captured	34	36
Recaptured[b]	8	17
Expected recaptures	12	13

[a]P^2= 1.32, *df* = 1, *P* = 0.25.
[b]P^2= 1.33, *df* = 1, *P* = 0.25.
Expected frequencies were based on sex ratio in captured population.

from different families can vary from mild to violent (Hodgdon 1978; Piechocki, 1977; Müller-Schwarze and Schulte, this volume). Beavers are known to vocalize, and to produce sounds by tail-slapping the water. Vocalizations are produced primarily by young beavers (Novakowski, 1969; Tevis, 1950) or during agonistic interactions (Hodgdon and Larson, 1973), and tail-slaps appear to communicate predominately with family members or potential predators (Hodgdon and Lancia, 1983). In the absence of direct contact, beavers are highly responsive to scent from conspecifics. The sources of scent are urine that washes through the castor sacs (i.e., called castor fluid or castoreum), and an oily fluid from the anal glands (Svendsen, 1978; Walro and Svendsen, 1982) Beavers deposit scent on objects around their site close to the water's edge, but *C. canadensis* and *C. fiber* differ somewhat in their mode of marking. North American beavers usually use mud from the pond bottom to create a mound (Hodgdon, 1978), while European beavers excavate small depressions or scratch together materials on land upon which to mark (Rosell and Bergan, 1998; Wilsson, 1971). Beavers will investigate and often overmark castoreum and anal gland secretions from foreign beavers placed on experimental scent mounds at their site (Hodgdon, 1978; Müller-Schwarze, 1992; Müller-Schwarze and Houlihan, 1991; Schulte, 1998; Sun and Müller-Schwarze, 1997; Svendsen and Huntsman, 1988). Castoreum is composed of hundreds of compounds (Tang, Webster and Müller-Schwarze, 1993, 1995). To date, no individual compounds or mixtures have evoked as strong a territorial response as whole castoreum (Müller-Schwarze and Houlihan, 1991; Schulte, Müller-Schwarze, Tang and Webster, 1994, 1995). Even the response to whole castoreum is variable, depending upon the identity of the donor (Schulte, 1998). For instance, adult males show little interest in their offspring's castoreum, but castoreum from strange adult males evokes strong territorial behavior. Anal gland secretion may serve a different communicative role from castoreum, acting more in kin recognition (Sun and Müller-Schwarze, 1997) and sex determination (Schulte, Müller-Schwarze and Sun, 1995), although some role in territorial defense may exist at least for European beaver (Rosell and Bergan, 1998).

The application of understanding the territorial nature of beavers and their reliance on chemical signals has been attempts to use natural scents to reduce colonization of potential beaver sites (Müller-Schwarze, 1995). The amount of scent marking by beavers increases with the density of sites and typically is highest during the time of spring dispersal (Houlihan, 1989; Rosell and Nolet, 1997; Rosell *et al.*, 1998; Svendsen, 1980a). Hence, scent marks may demarcate occupied territory to other beavers and be useful as a repellent. In studies to test this idea, sites marked with beaver castoreum or a mixture of castoreum and anal gland secretion were colonized significantly less than control sites (Müller-Schwarze and Heckman, 1980; Welsh and Müller-Schwarze, 1989). However, in a dense beaver population colonization was not significantly reduced at the scented sites (Müller-Schwarze, 1995). The use of intraspecific odors as repellents requires additional study and we will discuss some potential applications in section 4 on management.

Natural odors also may be useful in deterring foraging by beavers on specific trees. Odor signals from mammalian predators are a successful deterrent for feeding by some mammalian herbivores such as snowshoe hares (Sullivan, Nordstrom and Sullivan, 1985), pocket gophers (Sullivan, Crump, Wieser and Dixon, 1990) and mountain beavers (Nolte, Farley, Campbell, Epple and Mason, 1993) as well as beavers. Fecal odors from river otter (*Lutra canadensis* or *L. lutra*) and lynx (*Lynx lynx*) reduced foraging on aspen branches slightly more than odors from wolves (*Canis lupus*), domestic dogs (*C. familiaris*) or bears (*Ursus* spp.) in separate studies on *C. canadensis* (Engelhart and Müller-Schwarze, 1995) and C. *fiber* (Rosell and Czech, unpublished). In addition, chemicals from plants that beavers avoid or do not prefer have been effective in reducing the acceptability of aspen branches, a highly preferred species (Müller-Schwarze, Schulte, Sun, Müller-Schwarze and Müller-Schwarze, 1994 and Basey, this volume). As with intraspecific odors, compounds from predators and plants are variable in their effectiveness depending on such factors as the beaver population being studied and the time span over which repellency is measured.

4. BEAVER MANAGEMENT CONSIDERATIONS

Beavers settle in and modify riparian habitats, at times leading to conflicts with human interests. The increase in beaver numbers in North America earlier in the 20th century created fewer beaver-human conflicts than today. In many instances there were abundant areas to relocate unwanted individuals, much as is the case in parts of Europe today (Hartman, 1994, 1995, 1996; Macdonald, Tattersall, Brown and Balharry, 1995; Nolet and Rosell, 1998). Currently, there are few satisfactory habitats completely devoid of beavers in North America, and the translocation of individuals is no longer a viable solution in most cases. When translocation is a viable option timing is critical, and our lack of knowledge about transient beavers should be considered. For instance, translocated beavers are most likely to remain in the release area when moved during the fall (Novak, 1987; Gale Roberts, NRCS, personal communication). In addition, we need to understand the rate of expansion of current beaver populations into unsettled, yet apparently appropriate habitats. The translocation of beavers may simply delay problems in the source population. However, beavers may be an immediate benefit to areas and their relocation a very positive step (Stuebner, 1994; G. Roberts, personal communication). For instance, beavers have been used to stabilize the water table and increase forage for livestock in the western United States (Olson and Hubert, 1994). Yet, even in these areas beavers are a nuisance at times.

The most common form of control used with beavers is killing unwanted individuals or controlled harvesting of populations (see Hammerson, 1994 for a review of management issues). Typically this is performed by the use of traps, shooting, or livetrapping and then destroying the animals. Harvesting can be an effective means of population and local control as evidenced by the decimation of beaver numbers in the 1600–1800's in North America. Today, these methods are often less appealing or ineffective since the market for beaver products is greatly reduced from past centuries. Areas of conflict between humans and beavers are commonly on private lands, often requiring permits for the taking of beavers. Hence, one trapper can rarely set large numbers of traps without crossing numerous property lines. To the general public, beavers are viewed less as nuisance varmints and so killing is less palatable as a control method. However, killing of beavers still is prevalent. Approximately 20,000 beavers are harvested annually in Mississippi and some 80% of the carcasses are discarded; neither the fur nor meat is used (Dr. Dale Arner, personal communication). This seems to be a waste of a valuable resource, and in many situations, the beavers are more beneficial alive.

In terms of management, what are the alternatives to killing beavers? The use of fertility control devices or surgical procedures is an unlikely alternative at present (Arner, 1964; Gordon and Arner, 1976; Novak, 1987). These methods require the capture of live beavers, the up-front costs of the drug or surgical expertise, and the release back into the capture area. In most cases, the labor and costs for such procedures are exorbitant relative to the budget of state management programs. On a site-specific basis, such methods may be practical and valuable. When a family is removed from a site, the area becomes available for future immigrants. If the population density is high in the region, dispersal to this satisfactory site is likely and the local problem will recur. Maintenance of a mated pair that cannot produce offspring may be a reasonable alternative (Brooks *et al.*, 1980; Kennelly and Lyons, 1984). The adult pair may defend the site, preventing additional immigration, yet alter the landscape at a reduced rate than a growing family. An exciting area of future research will be to determine if beavers produce a natural reproductive suppressant under high-density conditions. Such a chemical signal may be very valuable in population control if it could be applied around active sites and lodges without requiring capture of individual beavers.

Natural predators can provide another form of population control for beavers. Gray wolves (*Canis lupus*) are the primary predators on beavers (Mech, 1966; Thurber and Peterson, 1993), although a number of other species including coyotes (*C. latrans*), bears (*Ursus* spp.) and even alligators (*Alligator mississippiensis*) prey on beavers (Arner *et al.*, 1981; Packard, 1940; Smith *et al.*, 1994). The presence of natural predators may slow or check the expansion of a beaver population. However, predators are unlikely to be of service for particular nuisance beavers as a short-term solution. While we advocate the maintenance of ecosystems with all their component parts, we realize that non-human predators will not solve all management dilemmas created by beavers.

Natural odors may have future application for controlling the activity of beavers. Marking areas with beaver scent can reduce colonization, but such practices are probably a short-term solution. As the beaver population in an area grows, beavers will investigate marked areas for territorial residents. Scent marking by beavers appears to act more in a scent-matching form of territoriality rather than as a scent fence (Sun and Muller-Schwarze, 1998). Hence, without a confrontation with a resident beaver, dispersing individuals will eventually settle in the unoccupied, yet marked site. Repellants and antifeedants, including synthetic compounds, may be effective on a local scale (Buech, 1985). Odor repellents discourage feeding on particular stands of trees, and may be more effective if other preferred forage is available nearby. However, for individual trees or small stands, fencing beavers out may be a more permanent and certain solution.

The most damaging consequence of beaver activity tends to be flooding which results from the building of dams and the blockage of culverts. In these situations, the foraging of beavers often is not a major problem. Fortunately, this situation has a fairly ready, inexpensive, and relatively permanent solution to allow the successful cohabitation of beavers and humans. The solution is water-flow control devices (Buech, 1985; Roblee, 1984; G. Roberts, personal communication). There are numerous types of devices that can be used in dams and around culverts. The basic principle is a long (3–6 m) pipe, commonly gated polyvinyl chloride (PVC) that is inserted into a beaver dam or extended from a culvert. The pipe may have T- or Y-shaped ends and is perforated by numerous holes (2–5 cm diameter). The holes create slow water seepage with little associated sound; both factors that can stimulate water stoppage activities by beavers. There are numerous variations on this design theme (Buech, 1985), but in general the devices work admirably, maintaining water levels acceptable for humans and beavers.

5. CONCLUSIONS

The adaptability of beavers is immense. North American beavers live from the subtropics to the boreal forests. Beavers can thrive in lakes, rivers, and streams as long as a year-round source of water is available. Beavers are relatively long-lived and can reproduce rapidly despite a single breeding season each year. Dispersing beavers are capable of travelling tens of kilometers to hundreds of kilometers. Given the opportunity, beavers can find suitable habitat even if initially it appears marginal. For beavers modify habitat — they are true ecosystem engineers. To some degree, this engineering can reduce environmental variation across their geographic range (analogous to the light and temperature controlled dwellings of humans). While beavers cannot alter these physical variables directly, they can mollify their impacts through habitat modification.

Beavers adapt to rapidly changing environmental conditions. Their response is to change the habitat back to a condition suitable for beavers. Trees are cut, dams are built and culverts are blocked. This is the crux of the dilemma for humans and the core of the beaver-human conflict. Beavers do not adjust to human activity by quietly filling in the unused environmental spaces. Rather, beavers attack the situation until it satisfies their needs. Humans can blow up dams, remove unwanted individuals and build structures such as roads and houses. However, if suitable or even marginal habitat persists, beavers will return and begin the process of modification to meet their needs. An understanding of individual behavior, ecological roles, life history characteristics, population dynamics and the historical impact of beavers in North America can assist humans in accepting this inevitability.

In general, the ecological benefits of beaver occupation in an area increase with time. When Europeans colonized North America in the 1600s, beavers were a ubiquitous and important part of the landscape. Undoubtedly, sympatric species evolved to the habitat modifications and cycles of change exacted by beavers. Beavers are not just a natural part of the riparian landscape, they are the architects of the system. If we redefined the word "nuisance" to a more tolerable level, then the costs of dealing with nuisance beavers would quickly drop. Given the flexibility of beaver behavior, perhaps we would be better to manage human activity, to use preventative measures to avoid problems with beavers and to reap the benefits of living with beavers.

ACKNOWLEDGMENTS

We thank the management of the various research sites where we have conducted our studies and the many students and volunteers who have assisted in this research. Our research has been supported by various sources including NSF Grant No. BNS-8819981 to DMS and a Theodore Roosevelt Fund grant to BAS. We appreciate the efforts and support of the organizing committee for the Euro-American Mammal Congress in Santiago de Compostela, Spain, and our home institutions, Providence College and SUNY-ESF. Our special thanks are extended to our co-participants in and attending colleagues at the symposium on beaver, especially the conveners, Peter E. Busher and Ryszard M. Dzięciołowski.

REFERENCES

Aleksiuk, M., 1968, Scent-mound communication, territoriality, and population regulation in beaver (*Castor canadensis* Kuhl), *J. Mamm.* **49:**759–762.

Allen, A. W., 1982, Habitat suitability index models: beaver, U.S. Dept. Int., Fish Wildl. Serv. FWS/OBS-82/10.30, p. 20.

Allred, M., 1981, The potential use of beaver population behavior in beaver resources management, *J. Idaho Acad. Sci.* **17:**14–24.

Allred, M., 1986, *Beaver Behavior: Architect of Fame and Bane*, NatureGraph Publ., Happy Camp, CA.

Arner, D. H., 1964, Research and a practical approach needed in management of beaver and beaver habitat in the southeastern United States, *Trans. North Amer. Wildl. Nat. Resour. Comm.* **29:**150–158.

Arner, D. H., Mason, C., and Perkins, C. J., 1981, Practicality of reducing beaver population through the release of alligators, in: *Worldwide Furbearers Conference Proceedings*, (J. A. Chapman and D. Pursley, eds.), pp. 1799–1805, Worldwide Furbearers Conference, Inc., Frostburg, MD.

Barnes, D. M., and Mallik, A. U., 1996, Use of woody plant in construction of beaver dams in northern Ontario, *Can. J. Zool.* **74:**1781–1786.

Barnes, D. M., and Mallik, A. U., 1997, Habitat factors influencing beaver dam establishment in a northern Ontario watershed, *J. Wildl. Manage.* **61:**1371–1377.

Barnes, W. J., and Dibble, E., 1988, The effects of beaver in riverbank forest succession, *Can. J. Bot.* **66:**40–44.

Basey, J. M., Jenkins, S. H., and Busher, P. E., 1988, Optimal central-place foraging by beavers: tree-size selection in relation to defensive chemicals of quaking aspen, *Oecologia* **76:** 278–282.

Beard, E. B., 1953, The importance of beaver in waterfowl management at Seney National Wildlife Refuge, *J. Wildl. Manage.* **17:**398–436.

Beier, P., and Barrett, R. H., 1987, Beaver habitat use and impact in the Truckee River basin, California, *J. Wildl. Manage.* **51:**794–799.

Bergerud, A. T., and Miller, D. R., 1977, Population dynamics of Newfoundland beaver, *Can. J. Zool.* **55:**1480–92

Bovet, J., 1994, Comment: Missed opportunities in the study of activity and body temperature of beaver (*Castor canadensis*), *Can. J. Zool.* **72:**567–569.

Boyce, M. S., 1981, Beaver life-history responses to exploitation, *J. Appl. Ecology* **18:**749–753.

Bradt, G. W., 1938, A study of beaver colonies in Michigan, *J. Mamm.* **19:**139–162.

Bradt, G. W., 1939, Breeding habits of beaver, *J. Mamm.* **20:**486–488.

Brady, C. A., and Svendsen, G. E., 1981, Social behaviour in a family of beaver, *Castor canadensis, Biol. Behav.* **6:**99–114.

Brenner, F. J., 1964, Reproduction of the beaver in Crawford County, Pennsylvania, *J. Wildl. Manage.* **28:**743–747.

Brooks, R. P., Fleming, M. W., and Kennelly, J. J., 1980, Beaver colony response to fertility control: evaluating a concept, *J. Wildl. Manage.* **44:**568–575.

Broschart, M. R., Johnston, C. A., and Naiman, R. J., 1989, Predicting beaver colony density in boreal landscapes, *J. Wildl. Manage.* **53:**929–934.

Brown, D. J., Hubert, W. A., and Anderson, S. H., 1996, Beaver pounds create wetland habitat for birds in mountains of southeastern Wyoming, *Wetlands* **16:**127–133.

Brown, M. K., 1979, Two old beavers from the Adirondacks, *New York Fish and Game J.* **26:**92.

Brown, M. K., and Parsons, G. R., 1979, Waterfowl production on beaver flowages in a part of northern New York, *New York Fish and Game J.* **26:**142–153.

Brown, M. K., and Parsons, G. R., 1982, Reliability of fall aerial censuses for locating active beaver colonies in northern New York, *New York Fish and Game J.* **29:**204–206.

Buech, R.R., 1985, Beaver in water impoundments: understanding a problem of water-level management, in: *Proceedings Water Impoundments for Wildlife: A Habitat Management Workshop*, (M. Dean Knighton, ed.), pp. 95–105, Aug. 31-Sept. 2, 1982, Bemidji, MN. USDA, FS, North Central Forest Experiment Station, St. Paul, MN, Gen. Tech Report NC-100, p. 136.

Buech, R. R., 1995, Sex difference in behavior of beavers living in near-boreal lake habitat, *Can. J. Zool.* **73:**2133–2143.

Busher, P. E., 1975, Movements and activities of beavers, Castor canadensis, on Sagehen Creek, California, M.A. thesis, San Francisco State univ., San Francisco, p. 86.

Busher, P. E., 1980, The population dynamics and behavior of beavers in the Sierra Nevada, Ph.D. dissert., Univ. Nevada, Reno, p. 147.

Busher, P. E., 1983, Interactions between beavers in a montane population in California, *Acta Zool. Fenn.* **174:**109–110.

Busher, P. E., 1987, Population parameters and family composition of beaver in California, *J. Mamm.* **68:**860–864.

Busher, P. E., and Jenkins, S. H., 1985, Behavioral patterns of a beaver family in California, *Biol. Behav.* **10:**41–54.

Busher, P. E., Warner, R. J., and Jenkins, S. H., 1983, Population density, colony composition, and local movements in two Sierra Nevadan beaver populations, *J. Mamm.* **64:**314–318.

Butts, W. L., 1988, Changes in local mosquito fauna following beaver (*Castor canadensis*) activity, *J. Am. Mosquito Cont. Assoc.* **2**:300–304.

Costanza, R., d'Agre, R., de Groot, R., Farber, S., Grasso, M., Hannon, B., Limburg, K., Naeem, S., O'Neill, R.V., Paruelo, J., Raskin, R.G., Sutton, P., and van den Belt, M., 1997, The value of the World's ecosystem services and natural capital, *Nature* **387**:253–60.

Davis, J. R. 1984, Movement and behavior patterns of beaver in the Piedmont of South Carolina, MS Thesis, Clemson University, Clemson, SC.

Dickinson, N. R. 1971, Aerial photographs as an aid in beaver management, *New York Fish and Game J.* **18:**57–61.

Dieter, C. D., and McCabe, T. R., 1989, Factors influencing beaver lodge-site selection on a prairie river, *Am. Midl. Nat.* **122:**408–411.

Dixon, B. R., Parenteau, M., Martineau, C., and Fournier, J., 1997, A comparison of conventional microscopy, immunofluorescence microscopy and flow cytometry in the detection of *Giardia lamblia* cysts in beaver fecal samples, *J. Immunol. Methods* **202:**27–33.

Doucet, C. M., Walton, R. A., and Fryxell, J. M., 1994, Perceptual cues used by beavers foraging on woody plants, *Anim. Behav.* **47:**1482–1484.

Easter-Pilcher, A., 1990, Cache size as an index to beaver colony size in northwestern Montana, *Wildl. Soc. Bull.* **18:**110–113.

Elsey, R. M., and Kinler, N., 1996, Range extension of the American beaver, *Castor canadensis*, in Louisiana, *The Southwestern Nat.* **41:**91–93.

Engelhart, A., and Müller-Schwarze, D., 1995, Responses of beaver (*Castor canadensis* Kuhl) to predator chemicals, *J. Chem. Ecol.* **21:**1349–1363.

Ermer, E. M., 1988, Managing beaver in New York State, *The Conservationist* **42:**36–39.

Ford, R., and Naiman, R. J., 1988, Alteration of carbon cycling by beaver: methane evasion rates from boreal forest streams and rivers, *Can. J. Zool.* **66:**529–533.

France, R. L., 1997, The importance of beaver lodges in structuring littoral communities in boreal headwater lakes, *Can. J. Zool.* **75:**1009–1013.

Fryxell, J. M. and Doucet, C. M., 1991, Provisioning time and central-place foraging in beavers, *Can. J. Zool.* **69:**1308–1313.

Gordon, K. L., and Arner, D. H., 1976, Preliminary study using chemosterilants for control of nuisance beaver, *Proc. Annu. Conf. Southeast Assoc. Fish and Wildl. Agencies* **30:**463–465.

Gore, J. A., and Baker, W. W., 1989, Beavers residing in caves in northern Florida, *J. Mamm.* **70:**677–678.

Gotie, R. F., and Jenks, D. L., 1984, Assessment of the use of wetlands inventory maps for determining potential beaver habitat, *New York Fish and Game J.* **31:**55–62.

Grover, A. M., and Baldassarre, G. A., 1995, Bird species richness within beaver ponds in south-central New York, *Wetlands* **15:**108–118.

Hammerson, G. A., 1994, Beaver (*Castor canadensis*) ecosystem alterations, management, and monitoring, *Natural Areas J.* **14:**44–57.

Hartman, G., 1994, Long-term population development of a reintroduced beaver (*Castor fiber*) population in Sweden, *Conserv. Biol.* **8:**713–717.

Hartman, G., 1995, Patterns of spread of a reintroduced beaver *Castor fiber* population in Sweden, *Wildl. Biol.* **1:**97–103.

Hartman, G., 1996, Habitat selection by a European beaver (*Castor fiber*) colonizing a boreal landscape, *J. Zool. Lond.* **240:**317–325.

Hay, K. G., 1958, Beaver census methods in the Rocky Mountain region, *J. Wildl. Manage.* **22:**395–402.

Henry, D. B., and Bookhout, T. A., 1969, Productivity of beavers in northeastern Ohio, *J. Wildl. Manage.* **33:**927–932.

Hilfiker, E. L., 1990, *Beavers: Water, Wildlife and History,* Windswept Press, Interlaken, NY.

Hill, E. P., 1982, Beaver (*Castor canadensis)*, in: *Wild Mammals of North America: Biology, Management, and Economics* (J. A. Chapman, and G. A. Feldhamer, eds.), pp. 256–281, John Hopkins Univ. Press, Baltimore, MD.

Hodgdon, H. E., 1978, Social dynamics and behavior within an unexploited beaver (*Castor canadensis*) population, PhD Thesis, Univ. Mass., Amherst, MA.

Hodgdon, H. E., and Lancia, R. A., 1983, Behavior of the North American beaver, *Castor canadensis, Acta Zool. Fennica* **174:**99–103.

Hodgdon, H. E., and Larson, J. S., 1973, Some sexual differences in behaviour within a colony of marked beavers (*Castor canadensis*), *Anim. Behav.* **21:**147–152.

Houlihan, P. W., 1989, Scent mounding by beaver (*Castor canadensis*): Functional and semiochemical aspects, MS Thesis, State Univ. of New York College of Environmental Science and Forestry, Syracuse, NY.

Howard, R. J., and Larson, J. S., 1985, A stream habitat classification system for beaver, *J. Wildl. Manage.* **49:**19–25.

Hudson, J. E., 1978, Canada's national mosquito? Mass-resting of *Anopheles earli* (Diptera: Culcidae) females in a beaver lodge in Alberta, *Can. Ent.* **110:**1345–1346.

Jenkins, S. H., and Busher, P. E., 1979, *Castor canadensis, Mamm. species* **120:**1–8

Johnson, C. E., 1927, The beaver in the Adirondacks: its economics and natural history, *Wild Life Bull.* **27:**501–641.

Johnson, D. R., and Chance, D. H., 1974, Presettlement overharvest of upper Columbia River beaver populations, *Can. J. Zool.* **52:**1519–1521.

Johnston, C. A., and Naiman, R. J., 1987, Boundary dynamics at the aquatic-terrestrial interface: the influence of beaver and geomorphology, *Landscape Ecol.* **1:**47–57.

Johnston, C. A., and Naiman, R. J., 1990a, Aquatic patch creation in relation to beaver population trends, *Ecology* **71:**1617–1621.

Johnston, C. A., and Naiman, R. J., 1990b, Browse selection by beaver: effects on riparian forest composition, *Can. J. For. Res.* **20:**1036–1043.

Johnston, C. A., and Naiman, R. J., 1990c, The use of a geographic information system to analyze long-term landscape alteration by beaver, *Landscape Ecol.* **4:**5–19.

Jones, C. G., Lawton, J. H., and Shachak, M., 1994, Organisms as ecosystem engineers, *Oikos* **69:**373–386.

Jones, C. G., Lawton, J. H., and Shachak, M., 1997, Positive and negative effects of organisms as physical ecosystem engineers, *Ecology* **78:**1946–1957.

Kennelly, J. J., and Lyons, P. J., 1984, Evaluation of induced sterility for beaver (*Castor canadensis*) management problems, *Proc. Eastern Wildl. Damage Control Conf.* **1:**169–175.

Lancia, R. A. 1979, Year-long activity patterns of radio-marked beaver (*Castor canadensis*) , Ph.D. dissert., Univ. Massachusetts, Amherst, p. 146.

Larson, J. S., and Gunson, J. R., 1983, Status of the beaver in North America, *Acta Zool. Fenn.* **174:**91–93.

Leege, T. A., 1968, Natural movements of beavers in southeastern Idaho, *J. Wildl. Manage.* **32:**973–976.

Leege, T. A., and Williams, R. M., 1967, Beaver productivity in Idaho, *J. Wildl. Manage.* **31:**326–332.

Libby, W. L., 1957, Observations on beaver movements in Alaska, *J. Mamm.* **38:**259.

Lyons, P. J., 1979, Productivity and population structure of western Massachusetts beavers, *Trans. Northeast Sec. Wildl. Soc. Fish Wildl. Conf.* **36:**176–187.

MacArthur, R. A., and Dyck, A. P., 1990, Aquatic thermoregulation of captive and free-ranging beavers (*Castor canadensis*), *Can. J. Zool.* **68:**2409–2416.

MacArthur, R. A., and Dyck, A. P., 1994, The search for a metabolic depression in the beaver: "lost opportunities" or red herrings? A reply to J. Bovet, *Can. J. Zool.* **72:**569–571.

McCall, T. C., Hodgman, T. P., Diefenbach, D. R., and Owen, R. B., 1996, Beaver populations and their relation to wetland habitat and breeding waterfowl in Maine, *Wetlands* **16:**163–172.

McComb, W. C., Sedell, J. R., and Buchholz, T. D., 1990, Dam-site selection by beavers in an eastern Oregon USA basin, *Great Basin Nat.* **50:**273–282.

Macdonald, D. W., Tattersall, F. H., Brown, E. D., and Balharry, D., 1995, Reintroducing the European beaver to Britain: nostalgic meddling or restoring biodiversity, *Mammal review* **25:**161–200.

McDowell, D. M., and Naiman, R. J., 1986, Structure and function of a benthic invertebrate stream community as influenced by beaver (*Castor canadensis*), *Oecologia* **68:**481–489.

McEvdy, C., and Jones, R., 1978, *Atlas of World Population History*, Penguin Books, New York.

Martinsen, G. D., Driebe, E. M., and Whitham, T. G., 1998, Indirect interactions mediated by changing plant chemistry: beaver browsing benefits beetles, *Ecology* **79:**192–200.

Mech, L. D., 1966, The wolves of Isle Royale, Fauna Natl. Parks U.S. Ser. 7, p. 210.

Mendin, D. E., and Clary, W. P., 1990, Bird populations in and adjacent to a beaver pond ecosystem in Idaho, U.S. Department of Agriculture, Forest Service, Intermountain Research Station, Ogden, UT, USA. Research Paper INT-432.

Molini, J. J., Lancia, R. A., Bishir, J., and Hodgdon, H. E., 1981, A stochastic model of beaver population growth, in: *Worldwide Furbearers Conference Proceedings*, (J.A. Chapman and D. Pursley, eds.), pp. 1215–1245, Worldwide Furbearers Conference, Inc., Frostburg, MD.

Müller-Schwarze, D., 1992., Castoreum of beaver (*Castor canadensis*): Function, chemistry and biological activity of its components, in: *Chemical Signals in Vertebrates 6* (R. L. Doty, and D. Müller-Schwarze, eds.) pp. 457–464, Plenum Press, New York.

Müller-Schwarze, D., 1995, Chemical repellents for beaver: new leads, in: *Chemical Signals in Vertebrates 7* (R. Apfelbach, D. Müller-Schwarze, and E. Weiler, eds.), pp. 479–483, Elsevier, Oxford.

Müller-Schwarze, D., 1996, Proactive management: avoiding conflict by knowing beaver behavior, in: *Living with Wildlife: The Biology and Sociology of Suburban Deer and Beaver* (A. N. Rowan, and J. C. Weer, eds.), pp 153–172, Tufts Center for Animals and Public Policy, N. Grafton, MA.

Müller-Schwarze, D., and Heckman, S., 1980, The social role of scent marking in beaver (*Castor canadensis*), *J. Chem. Ecol.* **6:**81–95.
Müller-Schwarze, D., and Houlihan, P. W., 1991, Pheromonal activity of single castoreum constituents in beaver, (*Castor canadensis*), *J. Chem. Ecol.* **17:**715–734.
Müller-Schwarze, D., Schulte, B. A., Sun, L, Müller-Schwarze, A., and Müller-Schwarze, C., 1994, Red maple (*Acer rubrum*) inhibits feeding by beaver (*Castor canadensis*), *J. Chem. Ecol.* **20:**2021–2034.
Naiman, R. J., and Melillo, J. M., 1984, Nitrogen budget of a subarctic stream altered by beaver (*Castor canadensis*), *Oecologia* **62:**150–155.
Naiman, R. J., Melillo, J. M., and Hobbie, J. E., 1986, Ecosystem alteration of boreal forest streams by beaver (*Castor canadensis*), *Ecology* **67:**1254–1269.
Naiman, R. J., Johnston, C. A., and Kelley, J. C., 1988, Alteration of North American streams by beaver, *BioScience* **38:**753–762.
Naiman, R. J., Pinay, G., Johnston, C. A., and Pastor, J., 1994, Beaver influences on the long-term biogeochemical characteristics of boreal forest drainage networks, *Ecology* **75:**905–921.
Neff, D. J., 1957, Ecological effects of beaver habitat abandonment in the Colorado Rockies, *J. Wildl. Manage.* **21:**80–84.
Nolet, B. A., and Rosell, F., 1998, Comeback of the beaver *Castor fiber*: an overview of old and new conservation problems, *Biol. Conserv.* **83:**165–173.
Nolet, B. A., Hoekstra, A., and Ottenheim, M. M., 1994, Selective foraging on woody species by the beaver *Castor fiber*, and its impact on a riparian willow forest, *Biol. Conserv.* **70:**117–128.
Nolte, D. L., Farly, J. P., Campbell, D. L., Epple, G. M., and Mason, J. R., 1993, Potential repellents to prevent mountain beaver damage, *Crop Protection* **12:**624–626.
Novak, M., 1977, Determining the average size and composition of beaver families, *J. Wildl. Manage.* **41:**751–754.
Novak, M., 1987, Beaver, in: *Wildlife Furbearer Management and Conservation in North America* (M. Novak, J. A. Baker, M. E. Obbard, and B. Malloch, eds.), pp. 283–312, Ashton-Potter Limited, Ontario Ministry of Natural Resources. Concord, Ont.
Novak, M., Baker, J., Obbard, M., and Mallock, B., eds., 1987. *Wildlife Furbearer Management and Conservation in North America*, Ashton-Potter Limited, Ontario Ministry of Natural Resources, Ontario.
Novakowski, N. S., 1969, The influence of vocalization on the behavior of beaver, *Castor canadensis* Kuhl, *Am. Midl. Nat.* **81:**198–204.
Nowak, R. M., 1991, *Walker's Mammals of the World*, 5th ed. John Hopkins Univ. Press, Baltimore, MD.
Olson, R., and Hubert, W. A., 1994, Beaver: water resources and riparian habitat manager, Univ. of Wyoming, Laramie, WY, p. 48.
Packard, F. M., 1940, Beaver killed by coyotes, *J. Mamm.* **21:**359–360.
Paine, R. T., 1966, Food web complexity and species diversity, *Am. Nat.* **100:**65–75.
Paine, R. T., 1969, The *Pisaster-Tegula* interaction: prey patches, predator food preference, and intertidal community structure, *Ecology* **50:**950–961.
Parsons, G., and Brown, M., 1978, An assessment of aerial photograph interpretation for recognizing potential beaver colony sites, *New York Fish and Game J.* **25:**175–177.
Patenaude, F., 1983, Care of the young in a family of wild beavers, *Castor canadensis*, *Acta Zool. Fennica* **174:**121–122.
Patenaude, F., and Bovet, J., 1983, Parturition and related behavior in wild American beavers (*Castor canadensis*), *Z. Säugetierkunde* **48:**136–145.
Payne, N. F., 1981, Accuracy of aerial censusing for beaver colonies in Newfoundland, *J. Wildl. Manage.* **45:**1014–1016.
Payne, N. F., 1982, Colony size, age, and sex structure of Newfoundland beaver, *J. Wildl. Manage.* **26:**272–278.
Payne N. F., 1984a, Mortality rates of beaver in Newfoundland, *J. Wildl. Manage.* **48:**117–126.
Payne N. F., 1984b, Reproductive rates of beaver in Newfoundland, *J. Wildl. Manage.* **48:**912–917.
Payne N. F., and Peterson, R. P., 1986, Trends in complaints of beaver damage in Wisconsin, *Wildl. Soc. Bull.* **14:**303–307.
Peterson, R. P., and Payne, N. F., 1986, Productivity, size, age, and sex structure of nuisance beaver colonies in Wisconsin, *J. Wildl. Manage.* **50:**265–268.
Piechocki, R., 1977, Ökologische Todesursachenforschung am Elbebiber (*Castor fiber albicus*), *Beitr. Jagd. Wildforsch.* **10:**332–341.
Radford, H. V., 1907, History of the Adirondack beaver (*Castor canadensis*, Kuhl.), Ann. Reports N.Y. State Forest, Fish and Game Comm. for 1904, 1905, 1906, pp. 389–418.
Renouf, R. N., 1972, Waterfowl utilization of beaver ponds in New Brunswick, *J. Wildl. Manage.* **36:**740–744.
Ricklefs, R. E., 1997, *The Economy of Nature*, 4th ed., W.H. Freeman and Co., New York.

Ringelman, J. K., 1991, Managing beaver to benefit waterfowl, Fish and Wildlife Leaflet 13.4.7, p. 7.

Robel, R. J., Fox, L. B., and Kemp, K. E., 1993, Relationship between habitat suitability index values and ground counts of beaver colonies in Kansas, *Wildl. Soc. Bull.* **21:**415–421.

Roberts, L. (ed.), 1998, *World Resources 1998–99: A Guide to the Global Environment*, The World Resources Institute, Oxford Univ. Press, NY.

Roblee, K. J., 1984, Use of corrugated plastic drainage tubing for controlling water levels at nuisance beaver sites, *New York Fish and Game J.* **31:**63–80.

Rosell, F., and Bergan, F., 1998, Free-ranging Eurasian beavers, *Castor fiber*, deposit anal gland secretion when scent marking, *Can. Field-Nat.* **112:** in press.

Rosell, F., and Kvinlaug, J. K., 1998, Methods for live-trapping beaver (*Castor* spp), *Fauna Norwegica*, in press.

Rosell, F., and Nolet, B. A., 1997, Factors affecting scent-marking behavior in the Eurasian beaver (*Castor fiber*), *J. Chem. Ecol.* **23:**673–689.

Rosell, F., Bergan, F., and Parker, H., 1998, Scent-marking in the Eurasian beaver (*Castor fiber*) as a means of territory defense, *J. Chem. Ecol.* **24:**207–219.

Ruedemann, R., and Schoonmaker, W. J., 1938, Beaver dams as geologic agents, *Science* **88:**523–525.

Schlosser, I. J., 1995, Dispersal, boundary processes, and trophic-level interactions in streams adjacent to beaver ponds, *Ecology* **76:**908–925.

Schulte, B. A., 1993, Chemical communication and ecology of the North American beaver (*Castor canadensis*), PhD Thesis, State Univ. of New York, College of Environmental Science and Forestry, Syracuse, NY.

Schulte, B. A., 1998, Scent marking and responses to male castor fluid by beavers, *J. Mamm.* **79:**191–203.

Schulte, B. A., Müller-Schwarze, D., Tang, R., and Webster, F. X., 1994, Beaver (*Castor canadensis*) responses to major phenolic and neutral compounds in castoreum, *J. Chem. Ecol.* **20:**3063–3081.

Schulte, B. A., Müller-Schwarze, D., Tang, R., and Webster, F. X., 1995, Bioactivity of beaver castoreum constituents using principle components analysis, *J. Chem. Ecol.* **21:**941–957.

Schulte, B. A., Müller-Schwarze, D., and Sun, L., 1995, Using anal gland secretion to determine sex in beaver, *J. Wildl. Manage.* **59:**614–618.

Slough, B. G., and Sadleir, R. M. F. S., 1977, A land capability classification system for beaver (*Castor canadensis* Kuhl), *Can. J. Zool.* **55:**1324–1335.

Smith, D. W., and Jenkins, S. H., 1997, Seasonal change in body mass and size of tail of northern beavers, *J. Mamm.* **78:**869–876.

Smith, D. W., and Peterson, R. O., 1991, Behavior of beaver in lakes with varying water levels in northern Minnesota, *Env. Manage.* **15:**395–401.

Smith, D. W., Trauba, D. R., Anderson, R. K., and Peterson, R. O., 1994, Black bear predation on beavers on an island in Lake Superior, *Am. Midl. Nat.* **132:**248–255.

Smith, M. E., Driscoll, C. T., Wyskowski, B. J., Brooks, C. M., and Cosentini, C. C., 1991, Modification of stream ecosystem structure and function by beaver (*Castor canadensis*) in the Adirondack Mountains, New York, *Can. J. Zool.* **69:**55–61.

Snodgrass, J. W., and Meffe, G. K., 1998, Influence of beavers on stream fish assemblages: effects of pond age and watershed position, *Ecology* **79:**928–942.

Stuebner, S., 1994, Bullish on beavers, *Natl. Wildl.* **32:**24–27.

Sullivan, T. P., Nordstrom, L. O., and Sullivan, D. S., 1985, Use of predator odors as repellents to reduce feeding damage by herbivores. I. Snowshoe hares (*Lepus americanus*), *J. Chem. Ecol.* **11:**903–919.

Sullivan, T. P., Crump, D. R., Wieser, H., and Dixon, E. A., 1990, Responses of pocket gophers (*Thomomys talpoides*) to an operation al application of synthetic semiochemicals of stoat (*Mustela erminea*), *J. Chem. Ecol.* **16:**941–949.

Sun, L., 1996, Chemical kin recognition in the beaver (*Castor canadensis*): Behavior, relatedness and information coding, PhD Thesis, State Univ. of New York, College of Environmental Science and Forestry, Syracuse, NY.

Sun, L., and Müller-Schwarze, D., 1998, Beaver response to recurrent alien scents: scent fence or scent match? *Anim. Behav.* **55:**1529–1536.

Sun, L., and Müller-Schwarze, D., 1997, Sibling recognition in the beaver: a field test for phenotype matching, *Anim. Behav.* **54:**493–502.

Svendsen, G. E., 1978, Castor and anal glands of the beaver (*Castor canadensis*), *J. Mamm.* **59:**618–620.

Svendsen, G. E., 1980a, Patterns of scent-mounding in a population of beaver (*Castor canadensis*), *J. Chem. Ecol.* **6:**133–147.

Svendsen, G. E., 1980b, Population parameters and colony composition of beaver (*Castor canadensis*) in southeast Ohio, *Am. Midl. Nat.* **104:**47–56.

Svendsen, G. E., 1989, Pair formation, duration of pair-bonds, and mate replacement in a population of beavers (*Castor canadensis*), *Can. J. Zool.* **67:**336–340.

Svendsen, G. E., and Huntsman, W. D., 1988, A field bioassay of beaver castoreum and some of its components, *Am. Midl. Nat.* **120:**144–149.

Swenson, J. E., Knapp, S. J., Martin, P. R., and Hinz, T. C., 1983, Reliability of aerial cache surveys to monitor beaver population trends on prairie rivers in Montana, *J. Wildl. Manage.* **47:**697–703.

Tang, R., Webster, F.X., and Müller-Schwarze, D., 1993, Phenolic compounds from male castoreum of the North American beaver (*Castor canadensis*), *J. Chem. Ecol.* **19:**1491–1500.

Tang, R., Webster, F.X., and Müller-Schwarze, D., 1995, Neutral compounds from male castoreum of North American beaver, *Castor canadensis*, *J. Chem. Ecol.* **21:**1745–1762.

Tevis, L. Jr., 1950, Summer behavior of a family of beavers in New York State, *J. Mamm.* **31:**40–65.

Thurber, J., and Peterson, R., 1993, Effects of population density and pack size on the foraging ecology of gray wolves, *J. Mamm.* **74:**879–889.

Townsend, J. E., 1953, Beaver ecology in western Montana with special reference to movements, *J. Mamm.* **34:**459–479.

Van Deelen, T. R., and Pletscher, D. H., 1996, Dispersal characteristics of two-year-old Beavers, *Castor canadensis*, in western Montana, *Can. Field-Nat.* **110:**318–321.

Walro, J. M., and Svendsen, G. E., 1982, Castor sacs and anal glands of the North American beaver (*Castor canadensis*): Their histology, development, and relationship to scent communication, *J. Chem. Ecol.* **8:**809–819

Welsh, R. G., and Müller-Schwarze, D., 1989, Experimental habitat scenting inhibits colonization by beaver, *Castor canadensis, J. Chem. Ecol.* **15:**887–893.

Wheatley, M., 1997a, Beaver, *Castor canadensis*, home range size and patterns of use in the taiga of southeastern Manitoba: I. Seasonal variation, Can. Field-Nat. **111:**204–210.

Wheatley, M., 1997b, Beaver, *Castor canadensis*, home range size and patterns of use in the taiga of southeastern Manitoba: II. Sex, age and family status, Can. Field-Nat. **111:**211–216.

Wheatley, M., 1997c, Beaver, *Castor canadensis*, home range size and patterns of use in the taiga of southeastern Manitoba: III. Habitat variation, Can. Field-Nat. **111:**217–222.

Wigley, T. B., Roberts, T. H., and Arner, D. H., 1983, Reproductive characteristics of beaver in Mississippi, *J. Wildl. Manage.* **47:**1171–1177.

Wilde, S. A., Youngberg, C. Y., and Hovind, J. H., 1950, Changes in composition of ground water, soil fertility, and forest growth produced by the construction and removal of beaver dams, *J. Wildl. Manage.* **14:**123–128.

Wilsson, L., 1971, Observations and experiments on the ethology of the European Beaver (*Castor fiber* L.), *Viltrevy* **8:**115–266.

Wolfe, M. L., and Chapman, J. A., 1987, Principles of furbearer management, in: *Wildlife Furbearer Management and Conservation in North America* (M. Novak, J. A. Baker, M. E. Obbard, and B. Malloch, eds.), pp. 101–112, Ashton-Potter Limited, Ontario Ministry of Natural Resources. Concord, Ont.

Woodward, D. K., 1983, Beaver management in the southeastern United States: a review and update, in: *First Eastern Wildlife Damage Control Conference Proceedings* (D.A. Decker, ed.), pp. 163–165, Cornell Coop. Ext., Ithaca, NY.

15

FORAGING BEHAVIOR OF BEAVERS (*CASTOR CANADENSIS*), PLANT SECONDARY COMPOUNDS, AND MANAGEMENT CONCERNS

John M. Basey

Department of EPOB
University of Colorado at Boulder
Boulder, Colorado 80309-0334

ABSTRACT

I examined how secondary metabolites in trees influenced food selection of beavers, whether selection patterns were consistent with recent theories of plant-defense investment, and whether a chemical extract could be used as a feeding deterrent. I used cafeteria-style experiments to examine beaver food preferences for six species, quaking aspen, willow, mountain alder, white fir, lodgepole pine, and Jeffrey pine; over four treatments, intact stems, stems with leaves removed, a methanol extract from bark, and a methanol extract from leaves. Quaking aspen was significantly preferred across all treatments, Jeffrey pine was significantly avoided across all treatments except the bark-extract treatment, and the bark extract of mountain alder was significantly avoided. Height growth rates and leaf lifespan of experimental trees were locally estimated and were not significantly different between the three deciduous species or the three coniferous species. Beavers selected significantly more deciduous than coniferous species across all treatments except the bark-extract treatment. A regression indicated a stronger relationship in selection of leaf extracts was present for leaf lifespans than height growth rate. In feeding experiments designed to examine the potential of an extract from Jeffrey pine as a repellent, beavers significantly avoided aspen stems painted with extracts from the bark and needles of Jeffrey pine when compared with controls. As a follow up, all aspen at two sites were either left untouched, painted with an extract from Jeffrey pine, or wrapped with chicken wire. The treatments significantly altered beaver food selection. These results are consistent with the hypothesis that food selection is influenced by secondary metabolites, that long-lived leaves are more protected from herbivory than short-lived leaves, and that chemical extracts from trees might be used as feeding deterrents. The greatest potential repellent ex-

Beaver Protection, Managment, and Utilization in Europe and North America, edited by Busher and Dzięciołowski.
Kluwer Academic / Plenum Publishers, New York, 1999.

tracts are from long-lived leaves from slow-growing trees, or bark from species specifically avoided by beavers.

1. INTRODUCTION

Compared with other large-vertebrate herbivores, North American beaver (*Castor canadensis*) and Eurasian beaver (*Castor fiber*) exhibit a disproportionate impact on ecosystem structure (Johnston and Naiman, 1990). One reason is that beavers are highly selective in their tree-felling activities, selecting between species and sizes of trees for food (Aldous, 1938; Jenkins, 1975; Jenkins, 1980; Pinkowski, 1983; Belovsky, 1984; McGinley and Whitham, 1985; Basey, Jenkins, and Busher, 1988; Fryxell and Doucet, 1991; Doucet and Fryxell, 1993; Fryxell, Vamosi, Walton, and Doucet, 1994; Nolet, Hoekstra, and Ottenheim, 1994; Basey and Jenkins, 1995). Through selective tree-felling, beavers can directly and/or indirectly change community structure in areas surrounding their ponds, and may be important in creating and maintaining critical habitat for various species (McGinley and Whittham, 1985; Johnston and Naiman, 1990; Nolet *et al.*, 1994; Martinsen, Driebe, and Whitham, 1998).

Selective tree-felling by beavers also causes conflicts with people (Nolet and Rosell, 1998), and current solutions to this conflict are not always the most desirable. A good example is the city of Reno, Nevada, USA, from 1989–1991. Downtown Reno is bisected by the Truckee River which is lined with large cottonwoods. During the winter of 1989, the cutting of cottonwoods by North American beavers in Reno increased to an extent that the public became alarmed. During the winter of 1990–1991, the problem was solved through a program of trapping and killing beavers along the Truckee River in Reno.

Since the potential impacts of selective tree-felling by beavers is so great, research concerning factors influencing tree-species selection and tree-felling behavior of beavers is vital for proper management of beavers. Of utmost importance is understanding why beavers select certain species, and the use of predictive models to implement management strategies. With this in mind, I examined food selection by beavers in relation to secondary metabolites and attempted to use generalized models to predict beaver feeding behavior.

Research has indicated that secondary metabolites can influence both the size, morphological type and species of trees selected for food by beavers. Basey *et al.* (1988) suggested that North American beavers in previously occupied stands of quaking aspen (*Populus tremuloides*) cut larger aspen compared with previously unoccupied stands most-likely due to high concentrations of a deterrent compound in the bark of smaller quaking aspen. Through cafeteria-style feeding experiments, Basey, Jenkins, and Miller (1990) demonstrated that North American beavers avoided extracts from the bark of juvenile-form sprouts of quaking aspen which arise from roots and stumps of aspen previously cut by beavers, and Müller-Schwarze, Shulte, Sun, Müller-Schwarze, and Müller-Schwarze (1994) demonstrated that North American beavers avoided extracts from the bark of red maple (*Acer rubrum*). Nolet *et al.* (1994) suggested that the complete avoidance of *Sambucus nigra* by Eurasian beavers in several studies is likely a consequence of the presence of a poisonous cyanogenic glycoside in the tree.

The influences of secondary metabolites on food preferences of beavers may also have important management applications. Extracts of secondary metabolites from tree species avoided by beavers have the potential to be used as repellents to alter tree-felling behavior of beavers (Müller-Schwarze *et al.*, 1994). Repellents could have a vital role in preventing damage of cultivated plants (trees and crops) by beavers but more research is needed (Nolet and Rosell, 1998).

It is possible that general models predicting investment in chemical defenses by plants can be used to predict food preferences of beavers for tree species, and to help define possible routes for chemical repellents. Two generalized hypotheses predicting relative investments of plants in defensive secondary metabolites are: (1) the resource-availability hypothesis (Coley, Bryant, and Chapin, 1985), and (2) the leaf-longevity hypothesis (Rhoades and Cates, 1976; McKey, 1979; Southwood, Brown, and Reader, 1986).

The resource-availability hypothesis is based on the idea that quantities of defenses in different plant species are largely determined by resource availability which in turn influences height-growth rates of trees. This hypothesis predicts that trees with intrinsically-fast growth rates will allocate more carbon to growth and less to chemical defenses compared with plants with intrinsically-slow growth rates (Coley *et al.*, 1985). Therefore, secondary metabolites from bark and/or leaves of slow-growing trees should be avoided by beavers when compared with those from relatively fast-growing trees.

The leaf-longevity hypothesis is based upon costs to the plant of producing a particular defense in relation to benefits of reduced herbivory (Figure 1). This hypothesis predicts that coniferous leaves living more than 5 years receive a greater relative benefit in carbon gain by investing more energy in the production of defenses than a deciduous leaf living for a fraction of a year. Therefore, secondary metabolites from relatively long-lived coniferous leaves should be avoided when compared with those from relatively short-lived deciduous leaves.

One problem with developing chemical deterrents for beavers in relation to these two hypotheses is the concept of mobile versus immobile defenses (Coley *et al.*, 1985). To the plant, immobile defenses such as lignin and mechanical defenses are initially costly to produce, but have low turnover rates, and often make food more difficult to process. Mobile defenses such as phenolic glycosides, monoterpenes, cyanogenic glycosides, and alkaloids, are relatively less-costly to produce, but must be continually synthesized by the plant, and are often toxic and cause feeding aversions (Coley *et al.*, 1985; Gulmon and Mooney, 1986).

In this study, I examined three primary questions. (1) What is the impact of mobile secondary metabolites on food preferences of woody-tree species by beavers? (2) Can broad generalized hypotheses of plant investment in mobile defenses in bark and leaves reliably predict beaver food preferences (e.g. the resource-availability hypothesis and the leaf-longevity hypothesis)? (3) Can mobile secondary metabolites from avoided species be used as deterrents/repellents and alter the relative selection of trees used for food by beavers?

2. METHODS

2.1. Study Sites

I conducted cafeteria-style feeding experiments at two sites. Site 1 was located along Franktown Creek, in the George Whittell Forest and Wildlife Area, Washoe County, Nevada, USA, approximately 39 km SSW of Reno, Nevada, at an elevation of 1,980 m in the Carson Range of the Sierra Nevada (Figure 2). The site was located in the middle of a mountain meadow with willow (*Salix* spp.) lining its edges and lodgepole pine (*Pinus contorta*) scattered throughout the area. The site contained several beaver dams. I observed at

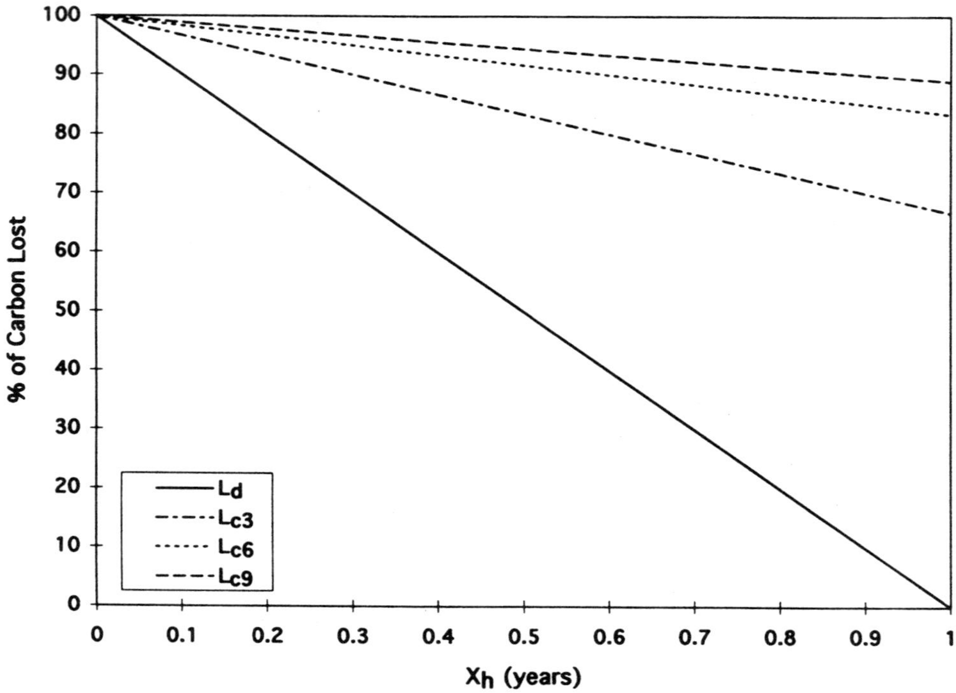

Figure 1. Percent of carbon lost by a plant when leaves living various numbers of years are prematurely eaten by herbivores. Percent of carbon lost was determined by the following mathematical formulation. The total net carbon gained by a plant leaf from photosynthesis (C_{net}) can be expressed mathematically as: $C_{net} = C_r x_l$; where C_r = net rate of carbon fixation (total carbon fixed - carbon invested per year), and x_l = leaf lifetime in years. The leaf lifetime of a tree depends upon life history traits (deciduous, x_l = 1 year; coniferous, x_l = several years) and mortality resulting from herbivory (which may be prevented for x_h years by investment in chemical defenses such that x_h ^{2} x_l). Let L_c = percent of carbon lost by a coniferous leaf living x_l years when eaten after x_h years compared to the total carbon which could have been gained in a full x_l years with no herbivory. Let L_d = percent of carbon lost by a series of deciduous leaves living 1 year each and eaten after x_h years (0.0 ^{2} x_h ^{2} 1.0) compared to the total carbon which would have been gained if all leaves lived 1 full year with no herbivory. L_c and L_d can be calculated by the equations: $L_c = [(x_l - x_h)/ x_l]100$; and $L_d = (1 - x_h)100$ (where 0.0 ^{2} x_h ^{2} 1.0 for deciduous leaves). L_{c3}, L_{c6}, and L_{c9} represent coniferous leaves living 3, 6, and 9 years respectively. This figure shows that % of carbon lost due to herbivory increases with leaf longevity. Therefore longer-lived leaves should receive a greater relative benefit in carbon gain by investing more energy in the production of defenses than shorter-lived leaves.

least six beavers in the summer of 1990 (four adults and two kits). Site 2 was located along Franktown Creek approximately 4 km south of site 1 in the Toiyabe National Forest. The area was dominated by quaking aspen (*Populus tremuloides*). Willow, mountain alder (*Alnus tenuifolia*), white fir (*Abies concolor*), lodgepole pine and Jeffrey pine (*Pinus jeffreyi*) were also present. The site contained several beaver dams. I observed at least six beavers in the summer of 1990 (four adults and two kits).

I conducted experiments testing the effectiveness of wrapping aspen with chicken wire and painting them with a deterrent extract at two sites, one of which was site 2 and the other I designate as site 3. Site 3 was located along Alder Creek (Figure 2), north of Truckee, Nevada County, California, approximately 4 km west of Highway 89. Vegetation was similar to that of site 2. The site was first occupied in mid summer 1990 and I observed at least 3 beavers at the site in spring 1991.

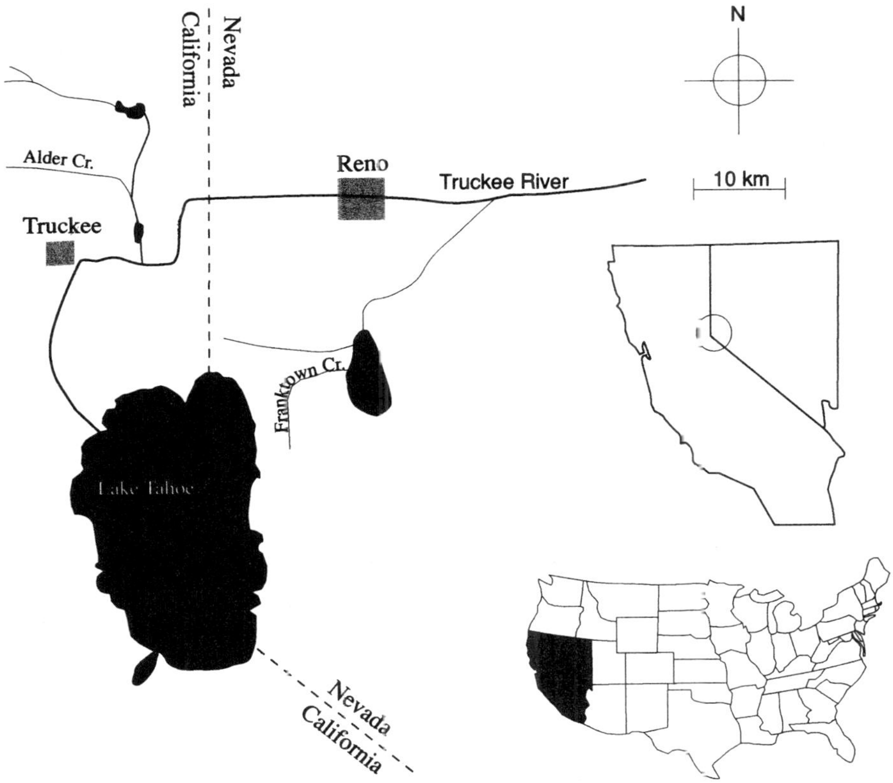

Figure 2. A map of the location of the study sites. Sites 1, 2, and 4 were located along Franktown Creek in Nevada. Site 3 was located along Alder Creek in California.

At site 4, I estimated height-growth rate and leaf longevity. Site 4 was located in a large light gap approximately 1/2 km north of site 1. A dirt road bisected the site. To one side, young Jeffrey pine, lodgepole pine, and white fir were available. On the other side of the road and closer to the stream, quaking aspen, mountain alder, and willow were available.

2.2. Height Growth Rate and Leaf Longevity

I selected the first 20 trees of each species (> 10 cm in diameter) I found at site 4. Tree age was estimated from growth rings of core samples taken approximately 20 cm above the ground. Tree height was determined with a tape measure. I selected ten branches from each tree to estimate leaf lifetime. For deciduous trees, I monitored the trees twice a week in the spring until > 50% of the buds broke, and in the fall, until > 50% of the leaves fell off. For white fir, a stem with secondary branching grows annually from the primary branch (Fowells, 1965). Therefore, I counted the number of secondary branches from the tip of the primary branch until the point on the primary branch where no living needles were present. For Jeffrey and lodgepole pine, secondary branching does not occur annually from the primary branch, but a gap exists between clumps of needles from one year to the next (Fowells, 1965). Therefore, I counted the number of gaps between clumps of needles.

2.3. Feeding Experiments

Cafeteria-style feeding experiments were conducted at site 1 from August to November 1990 and 1991, and at site 2 from August to November 1990. Experiments were not continued at site 2 in 1991 because many of the beavers had been trapped and killed earlier in the summer of 1991. Next to the creek, I cleared 2 by 8 m plots of vegetation and contours to give beavers equal access to experimental stems. At each site two wooden boards, each with 21 holes 1.3 cm in diameter and 15 cm apart, were buried end to end in the soil and their upper surfaces covered with soil. Experimental tree branches were placed into the holes, and soil was packed around each so that a beaver had to cut a stem to remove it. Beavers were given a choice of six species of trees: quaking aspen, willow, mountain alder, white fir, lodgepole pine, and Jeffrey pine. All branches were within the range of 0.9 to 1.2 cm in basal diameter and collected from trees < 10 cm in diameter. Each trial consisted of 42 branches equally divided among species and randomly positioned. Branches were examined daily for any signs of beaver activity (tracks, cut branches, etc.). All cut branches were replaced with fresh ones, and an equal number of branches of other species were also replaced. For example, if 3 quaking aspen branches were the only branches selected during a trial, I would replace the 3 quaking aspen branches and 3 randomly selected branches of each species. All branches were completely replaced every four days. Six different types of experiments were run following this general procedure. Beavers were given the choice of the folllowing.

1. Intact branches of the six species.
2. Branches of the six species with leaves removed.
3. Branches of quaking aspen with 95% methanol extracts from leaves of the six species.
4. Branches of quaking aspen with 95% methanol extracts from the bark of the six species.
5. Branches of quaking aspen with 95% methanol extracts from the bark and leaves of Jeffrey pine and control branches of quaking aspen with 95% methanol (sites 1 and 2, 1990).
6. Branches of quaking aspen with 95% methanol extracts from the bark and leaves of Jeffrey pine previously soaked in water, and control branches of quaking aspen with 95% methanol (sites 1 and 2, 1990).

Methanol-extract experiments for 6-species comparisons (treatments 1–4) were carried out in the following manner. At the beginning of a given 4-day trial, I collected branches of each species (40 branches/species in 1990 and 20 branches/species in 1991), stripped bark and leaves from each stem, finely cut them with scissors, added them separately by species and type (bark or leaves) to 600 ml of cold 95% methanol (300 ml in 1991) and left them for 24 hours. I then sprayed extracts from each species onto separate groups of 40 aspen branches in 1991 and 20 aspen branches in 1990, then left them to dry. During the four day trial, all cut branches were replaced with stems which were prepared on day 1 of the trial.

Methanol-extract experiments for Jeffrey pine (treatments 5 and 6) were carried out with the above procedure, but a few changes were made. For treatment 5, I collected 21 stems of Jeffrey pine and prepared samples of bark/leaves. I added the samples to 200 ml of cold 95% methanol and left them for 1 hour. I painted extracts onto 21 aspen stems, and left them to dry for 3 hours. Simultaneously, I painted 200 ml of 95% methanol onto 21 aspen stems to act as a control. For treatment 6, I used similar methods to those of treatment 5, but prior to placing bark and leaf samples of Jeffrey pine into 95% methanol, I placed them in 200 ml of water for 1 hour, and repeated this step two more times, discarding the solution each time.

2.4. Fiber and Lignin Analysis

To sample quantities of fiber and lignin in the experimental trees, I collected 20 branches of each species, stripped the bark and finely cut it with scissors, thoroughly mixed the bark, and freeze dried it. I then used 10 g from the mixture from each species. I collected the samples in October 1991. The 10 g samples of all six species were analyzed by the Animal Nutrition Laboratory at the University of Nevada, Reno. Acid detergent fiber, crude fiber, and lignin were analyzed following procedures outlined by Jones (1984).

2.5. Deterrent/Repellent Experiments

Experiments were conducted at sites 2 and 3 from 15 October through 15 December 1991. At each site, all aspen > 5.0 cm in diameter and < 20.0 m from the pond were randomly divided into 3 treatment groups: (1) aspen with chicken wire wrapped around their bases to a height of 75 cm from the ground; (2) aspen with a propanol extract from the bark and leaves of Jeffrey pine painted on their bases to a height of 75 cm above the ground; and (3) aspen with no protection from beavers. During the experimental set up, I recorded tree size and distance from the shore. Since the Environmental Protection Agency (EPA) would not allow the use of methanol for this series of experiments, I used propanol to extract deterrents from the bark and leaves of Jeffrey pine. To extract deterrents from Jeffrey pine, I stripped 1,000 g of bark and leaves from several Jeffrey pine trees < 4.0 cm in diameter and combined them with 2.0 l of propanol, and let them soak for 24 hours. At each site, I poured the extract into a pressurized spray can and sprayed it onto the bases of the experimental aspen. Following the experimental set up, I recorded the number of aspen in each treatment cut by beavers every week until 15 November and every 2 weeks until 15 December.

3. RESULTS

3.1. Height Growth Rate and Leaf Longevity

Estimates of mean height growth rate for the 6 species indicated clear differences between deciduous and coniferous species (Table 1). A sequential Bonferroni test (Rice 1989) examining all pair-wise combinations of tree species revealed a significant difference between deciduous species (quaking aspen—QA, willow—W, and mountain alder—MA) and coniferous species (white fir—WF, lodgepole pine—LP, and Jeffrey pine—JP), but no significant differences within groups of deciduous and coniferous species (Table 2).

Leaf lifespans were approximately 5 months for all three deciduous species. Fifty percent of the buds were broken on selected trees of all 3 species between 13 May and 20 May 1991, and leaves fell off between 18 October and 8 November.

Mean leaf lifespan was different between deciduous and coniferous species and among coniferous species (Table 1). An ANOVA with coniferous species as the independent variable and leaf longevity (years) as the dependent variable demonstrated that

Table 1. Estimates of mean height growth rate and one standard deviation (n = 20) of the six tree species used in the feeding experiments[a]

Species	Height growth rate (m/yr)	Leaf lifespan
Mountain alder	0.32 ± 0.12	0.42
Willow	0.31 ± 0.12	0.42
Quaking aspen	0.26 ± 0.09	0.42
Lodgepole pine	0.16 ± 0.04	6.68 ± 1.05
White fir	0.13 ± 0.02	6.89 ± 1.30
Jeffrey pine	0.11 ± 0.01	5.53 ± 0.86

[a]For deciduous species (mountain alder, willow, and quaking aspen), buds broke and leaves were lost on all trees in a relatively short time so no standard deviation is estimated.

longevity was significantly different between groups ($F = 7.56$, *d.f.* = 2, 547, $0.001 < P < 0.01$). A sequential Bonferroni test examining pair-wise combinations revealed no significant differences between leaf longevity of white fir and lodgepole pine ($P > 0.05$), but significant differences were present between Jeffrey pine and lodgepole pine, and Jeffrey pine and white fir ($P < 0.01$).

3.2. Feeding Experiments

The first set of feeding experiments consisted of four treatments: intact branches, branches with leaves removed, aspen branches sprayed with methanol extracts from bark, aspen branches sprayed with methanol extracts from leaves. I used a replicated goodness of fit test (Sokal and Rohlf 1981, pp. 722–724), where site 1–1990, site 2–1990, and site 1–1991 were treated as separate replicates and the expected frequency of selection of each species was 1/6. The test showed that selection of branches by species was significantly different from random for intact branches, for branches with leaves removed, and for leaf extracts, but selection of bark extracts was significantly different only at site 1 in 1991 (Table 3). The replicated goodness of fit test also indicated that no significant differences in selection of species were present at any of the sites between intact branches and branches with leaves removed (site 1, 1990; $G = 10.45$, *d.f.* = 5, $P > 0.05$; site 2, 1990; $G = 2.59$, *d.f.* = 5, $P > 0.05$; site 1, 1991; $G = 7.43$, *d.f.* = 5, $P > 0.05$; Figure 3).

I used a *G*-test of goodness of fit with William's correction for continuity (Sokal and Rohlf 1981, pp. 705–708) to examine selection by species within each treatment for pooled data from all sites. The expected frequency of selection of a single species was 1/6

Table 2. Results of sequential Bonferroni test (Rice 1981) comparing height growth rates of six species: W = willow, MA = mountain alder, QA = quaking aspen, WF = white fir, LP = lodgepole pine, and JP = Jeffrey pine

Species	W	MA	QA	WF	LP
JP	Y	Y	Y	N	N
LP	Y	Y	Y	N	
WF	Y	Y	Y		
QA	N	N			
MA	N				

Y = a significant difference ($0.001 < P < 0.01$). N = no significant difference ($P > 0.05$).

Table 3. Significance (G) of branch selection by beavers of species during cafeteria-style feeding experiments from four treatment groups: (1) intact branches, (2) branches with leaves removed, (3) aspen branches sprayed with methanol extracts from bark, (4) aspen branches sprayed with methanol extracts from leaves

Test	d.f.	Intact	No leaves	Bark ext.	Leaf ext.
Pooled	5	303.79***	67.61***	12.30*	26.31***
Het.	10	25.82**	8.63	3.1	5.24
Total	5	329.61***	76.25***	15.4	31.55**
Sites					
1 (1990)	5	145.81***	10.26**	1.7	7.85
2 (1990)	5	81.11***	27.54***	3.19	10.22**
1 (1991)	5	102.70***	38.45***	10.52**	13.47***

Pooled refers to data pooled between the three sites, and Het. refers to heterogeneity between sites. Significance levels are indicated as follows: ***, $P < 0.001$; **, $0.001 < P < 0.01$; no stars indicate $P > 0.05$.

and that for the other 5 species was 5/6. For intact stems, beavers selected quaking aspen (QA) and willow (W) in significantly greater numbers than predicted by random selection, and they selected white fir (WF), mountain alder (M), lodgepole pine (LP) and Jeffrey pine (JP) in significantly lower numbers than predicted by random selection (Figure 4). For bark extracts, beavers selected QA significantly more than random, but selected MA significantly less than random (Figure 5). For leaf extracts, beavers selected QA significantly more than random, and WF and JP significantly less than random (Figure 6).

I combined data from feeding experiments into two groups: deciduous species and coniferous. I used a G-test with William's correction for continuity to examine between the two groups. The expected frequency of selection was 1/2 for coniferous species and 1/2 for deciduous species. Beavers selected significantly more stems associated with deciduous than coniferous species for intact stems ($G_{adj} = 115.8$, $d.f. = 1$, $P < 0.001$) and leaf extracts ($G_{adj} = 22.94$, $d.f. = 1$, $P < 0.001$), but no significant differences were present for bark extracts ($G_{adj} = 2.07$, $d.f. = 1$, $P > 0.05$; Figure 7).

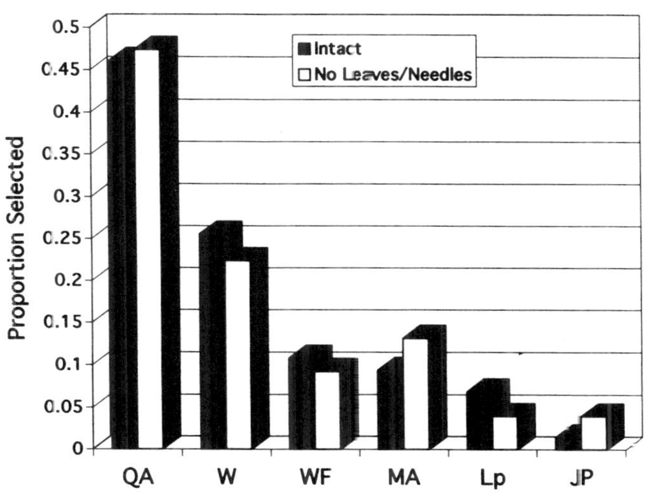

Figure 3. Selection of intact branches with bark and leaves/needles compared to selection of branches with leaves/needles removed. Data from all sites were combined. No significant differences of species between the two treatments were present ($P > 0.05$).

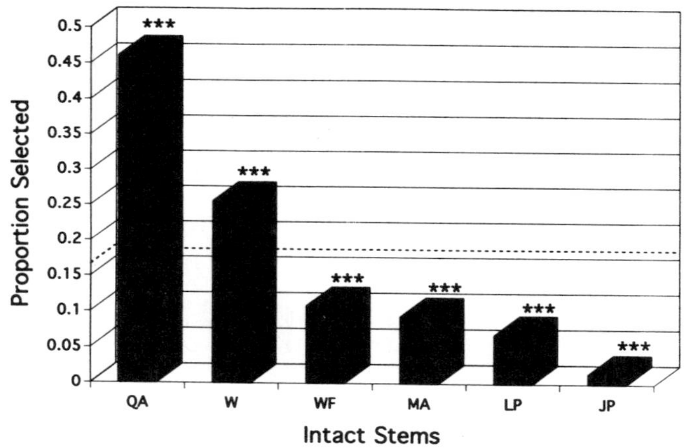

Figure 4. Selection between species of intact stems in the cafeteria-style feeding experiments. Available species were: quaking aspen (QA), willow (W), white fir (WF), mountain alder (M), lodgepole pine (LP) and Jeffrey pine (JP). Selection of each species was compared to random selection with a goodness of fit test. Random selection is indicated with the dashed line. Selection was significantly different than random for all species (***, $P < 0.001$).

I examined the relationship between height growth rate and proportion of stems of each species selected by treatment with a regression analysis. The test showed no significant relationships were present for intact stems (Figure 8, $F = 3.94$, *d.f.* = 1, 16, $P = 0.064$, $R^2 = 0.20$) and bark extracts (Figure 9, $F = 2.41$, *d.f.* = 1, 16, $P = 0.14$, $R^2 = 0.13$), but there was a significant positive regression for leaf extracts (Figure 10, $F = 20.82$, *d.f.* = 1, 16, $P = 0.0003$, $R^2 = 0.57$). There was also a significant negative regression for leaf lifespan in relation to leaf extracts (Figure 11, $F = 42.78$, *d.f.* = 1, 16, $P > 0.0001$, $R^2 = 0.73$).

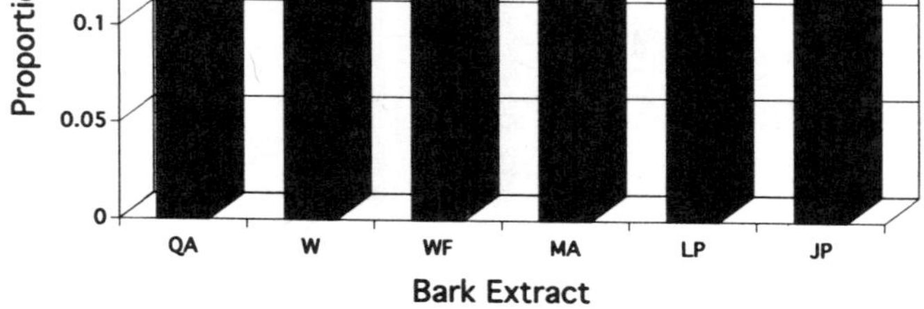

Figure 5. Selection during cafeteria-style feeding experiments of aspen stems painted with a methanol extract from the bark of six species: quaking aspen (QA), willow (W), white fir (WF), mountain alder (M), lodgepole pine (LP) and Jeffrey pine (JP). Data were combined from all sites. Selection of each species was compared to random selection with a goodness of fit test. Random selection is indicated with a dashed line. Significance levels are indicated as follows: *, $0.01 < P < 0.05$; no stars indicate $P > 0.05$.

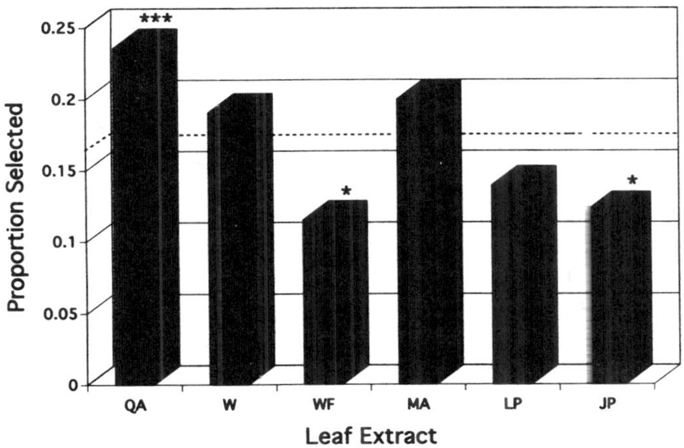

Figure 6. Selection during cafeteria-style feeding experiments of aspen stems painted with a methanol extract from the leaves of six species: quaking aspen (QA), willow (W), white fir (WF) mountain alder (M), lodgepole pine (LP) and Jeffrey pine (JP). Data were combined from all sites. Selection of each species was compared to random selection with a goodness of fit test. Random selection is indicated with a dashed line. Significance levels are indicated as follows: ***, $P < 0.001$; *, $0.01 < P < 0.05$; no stars indicate $P > 0.05$.

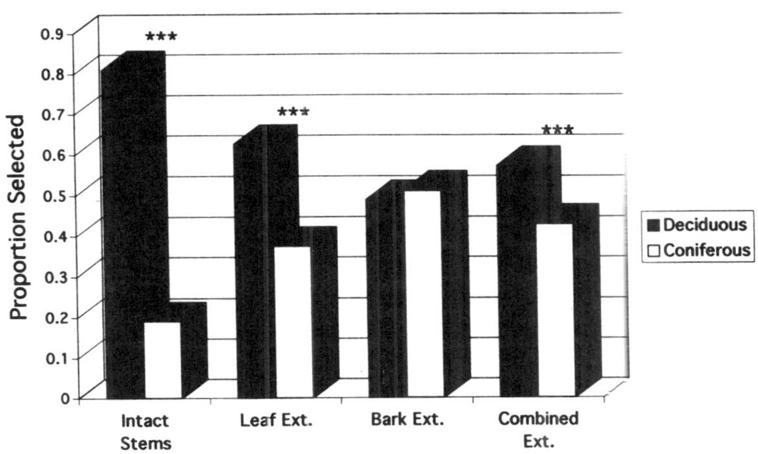

Figure 7. A comparison of the selection of experimental stems selected during cafeteria-style feeding experiments grouped as either deciduous species (mountain alder, willow, and quaking aspen) or coniferous species (white fir, lodgepole pine, and Jeffrey pine). Data were combined for all sites. Significance levels are indicated as follows: ***, $P < 0.001$; no stars indicate $P > 0.05$.

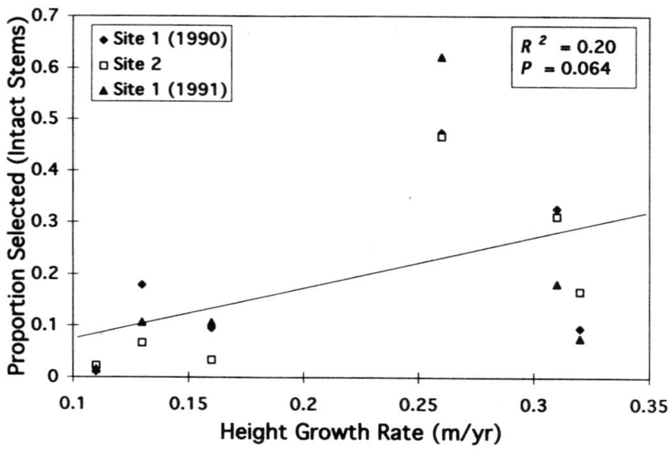

Figure 8. Proportion of intact stems of six species selected at the three sites during cafeteria-style feeding experiments in relation to height growth rate. No significant relationship is present between proportion of stems selected and height growth rate.

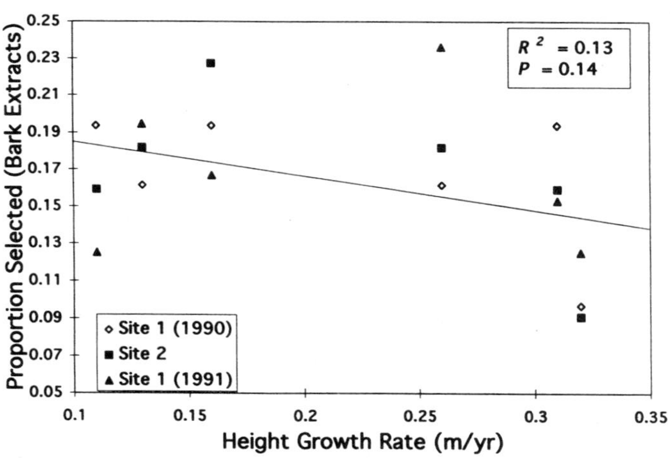

Figure 9. Selection of aspen stems painted with bark extracts during cafeteria-style feeding experiments in relation to height growth rates of the six experimental species. No significant relationship is present between proportion of stems selected and height growth rate.

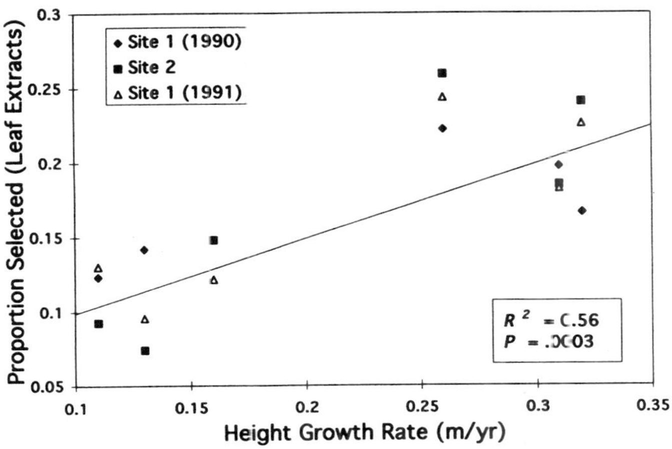

Figure 10. Selection of aspen stems painted with leaf extracts during cafeteria-style feeding experiments in relation to height growth rates of the six experimental species. A significant relationship is present between proportion of stems selected and height growth rate.

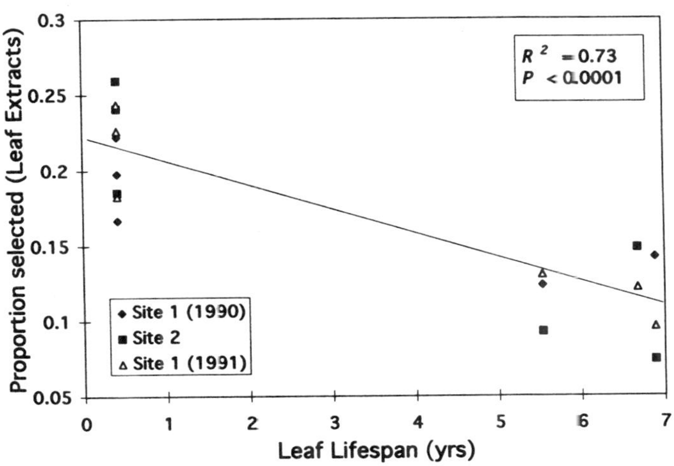

Figure 11. Selection of aspen stems painted with leaf extracts during cafeteria-style feeding experiments in relation to leaf lifespans of the six experimental species. A significant relationship is present between proportion of stems selected and height growth rate.

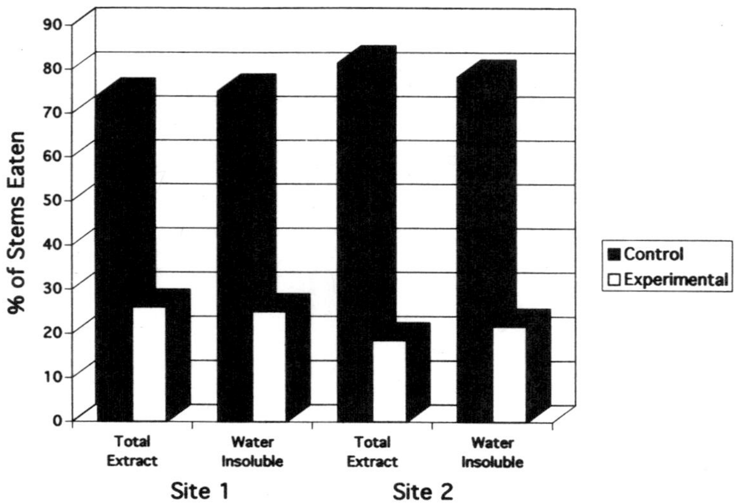

Figure 12. Percent of stems selected by beavers in extract painting experiments. Control stems were painted with plain 95% methanol and experimental stems were painted with a 95% methanol extract from Jeffrey pine. One experiment was conducted where the total methanol extract from Jeffrey pine bark was used, and a second where Jeffrey pine bark and leaves were washed with water prior to methanol extraction. Significance levels are indicated as follows: ***, $P < 0.001$.

3.3. Nutrition Analysis

The nutritional analysis indicated that quaking aspen had lower bark concentrations of acid detergent fiber (ADF), lignin, and cellulose than the other five species (Table 4). The analysis also revealed a relatively high lignin content in the bark of Mountain Alder and Jeffrey pine (Table 4).

3.4. Deterrent Experiments

In cafeteria-style feeding experiments, selection of aspen stems painted with a 95% methanol extract from the bark and leaves of Jeffrey pine (experimental) or with straight 95% methanol (control) revealed that beavers selected experimental stems significantly less often than control stems (Figure 12). When water was used to extract chemicals from Jeffrey pine bark and leaves prior to extraction with 95% methanol, beaver selection patterns between experimental and control aspen did not change (Figure 12).

An analysis of variance (ANOVA) showed no significant differences in diameter or distance from pond of experimental aspen between treatment groups (chicken wire, extract, unprotected) at either site (site 2—diameter: $F = 0.376$, *d.f.* = 2, 160, $P > 0.05$; distance: $F = 0.376$, *d.f.* = 2, 160, $P > 0.05$; site 3—diameter: $F = 0.182$, *d.f.* = 2, 144, $P > 0.05$; distance: $F = 0.032$, *d.f.* = 2, 144, $P > 0.05$).

Selection of quaking aspen by beavers was altered from random selection by the treatments (Figure 13). I combined data for sites 2 and 3 and used a *G*-test with William's correction for continuity to examine between treatments. The expected frequency of selection of an aspen in each of the three treatments was set at 1/3. Beavers selected untreated aspen significantly more often than predicted by random selection ($G_{adj} = 8.06$, *d.f.* = 1, $P < 0.001$), selection of quaking aspen painted with extract from Jeffrey pine was not sig-

Figure 13. Number of quaking aspen of three treatments (untouched, sprayed with a propanol extract from the bark and leaves of Jeffrey pine, and wrapped with chicken wire) felled at sites 2 and 3 during deterrent experiments. The number selected was compared with random selection. Significance levels are indicated as follows: ***, $\underline{P} < 0.001$; no stars indicate $\underline{P} > 0.05$.

nificantly different than predicted by random selection (G_{adj} = 9.99, *d.f.* = 1, $P < 0.001$), and they selected quaking aspen wrapped in chicken wire significantly less often than predicted by random selection (G_{adj} =7.87, *d.f.* = 1, $P < 0.001$).

4. DISCUSSION

Since selection of intact branches during the cafeteria-style feeding experiments did not significantly differ from branches with leaves removed, beavers appear to have used characteristics of the bark to make selections (Figure 3). In contrast, observations indicate that beavers selected intact stems by species based upon combined characteristics of leaves and bark. Two examples are mountain alder and white fir. During cafeteria-style feeding experiments of intact stems, I commonly found experimental stems of mountain alder in the water with the leaves removed but the bark intact, and for white fir, bark on the main stem was stripped while side branches with dense needles were left intact. Both of these selection patterns are consistent with cafeteria-style feeding experiments using

Table 4. A comparison of concentrations of compounds which reduce digestion efficiency in the bark of the six experimental species

Defense	QA	W	WF	MA	LP	JP
ADF	27.81	36.02	36.00	34.04	38.02	35.14
Cellulose	12.77	18.13	18.85	15.55	21.70	13.57
Lignin	12.26	17.89	17.15	21.27	16.32	21.57
Total	52.84	72.04	72.00	70.86	76.04	70.28

Values presented are percent of oven dry weight. In the table QA is quaking aspen, W is willow, WF is white fir, MA is mountain alder, LP is lodgepole pine, and ADF is acid detergent fiber.

treatments of bark extracts (Figure 5), and leaf extracts (Figure 6). Beavers avoided bark extracts of mountain alder, whereas they avoided leaf extracts of white fir. In addition at site 1, I commonly found young white fir, approximately 5 cm in diameter, with branches cut from the main trunk and left intact, and the bark from the main trunk stripped and eaten; thus, indicating that beavers avoided the leaves of white fir, but readily ate the bark.

Height growth rates of experimental trees did not account for selection of intact stems by beavers (Figure 8). The key exception was mountain alder which was the fastest growing of the 6 species but was selected significantly less than random (Figure 8). Part of the reason for the avoidance of intact mountain alder was the presence of mobile secondary metabolites in its bark (Figure 5). Other than mountain alder and quaking aspen, bark extracts did not account for selection of species in cafeteria-style feeding experiments with intact stems (Figure 5), and there is no evidence that quantities of mobile defenses in the bark increase with height-growth rate (Figure 8).

It is possible that experimental trees invested relatively more energy into immobile defenses in the bark than mobile defenses and selection of experimental trees was determined by immobile defenses such as lignin and fiber. This is supported by the nutritional analysis in which quaking aspen had lower bark concentrations of potential immobile defenses than the other five species (Table 4). Food preferences by beavers for willow during the feeding experiments, however, are not explained by relatively low quantities of immobile defenses (Table 4), but may be explained when leaf preferences are factored in. Other research supports this contention and has demonstrated that preferences for tree species utilized for food by North American beavers are influenced by retention times of food in the North American beavers which is likely determined by lignin and fiber content (Doucet and Fryxell, 1993; Fryxell *et al.*; 1994).

The resource availability hypothesis and leaf longevity hypothesis both predict that beavers should prefer deciduous over coniferous species used in the feeding experiments. Selection of aspen painted with leaf extracts was consistent with both hypotheses (i.e. beavers significantly preferred deciduous to coniferous species, Figure 7). The regression analysis revealed a much stronger relationship between selection of leaf extracts and leaf longevity (Figure 11), than selection of leaf extracts and height growth rate (Figure 10). This is consistent with past research concerning these two hypotheses. Several studies in relation to the resource availability hypothesis have had results inconsistent with predictions of the hypothesis (Newberry and de Foresta, 1985; Denslow, Vitousek, and Schultz, 1987; McCanny, Keddy, Arnason, Gaudet, Moore, and Shipley, 1990; Denslow, Schultz, and Vitousek, 1990; but see Basey and Jenkins, 1993); whereas, the leaf-longevity hypothesis has been consistently supported (Coley and Barone, 1996).

Results of the deterrent experiments demonstrated that an extract from non-preferred species may be a successful beaver repellent but more research is needed. Cafeteria-style feeding experiments demonstrated that the methanol extract from the bark and leaves of Jeffrey pine could act as a repellent when painted onto aspen stems with unpainted aspen available (Figure 12). These experiments also demonstrated that water insoluble components from the bark and leaf extract of Jeffrey pine could also act as a deterrent; thus, indicating the potential for a deterrent to withstand rainy and snowy conditions. In these experiments, samples were continuously replaced and encounter rates were kept relatively consistent, which does not mirror the use of a repellent in the field. Field tests comparing a more realistic application of repellents showed that selection of quaking aspen at two natural sites was altered by the use of a propanol extract of Jeffrey pine and chicken wire in relation to untouched trees (Figure 12). The results indicate that chicken wire was more effective in reducing cutting by beavers than the propanol extract, but the extracts in general have some potential as repellents. These experiments, however, had a small sample size and further testing is required, especially tests with an extract as the sole treatment. In addition, beavers typically occupy forests with mixed species, and the ability of a chemical repellent to protect a preferred species over a non-preferred species, and all species in a given area must be examined. Swihart and Conover 1991 showed that when food types varying in preference to woodchucks (*Marmota monax*) were all treated with a repellent, the repellent reduced herbivory in the less palatable species, and did not reduce herbivory in the more palatable species.

Results of the cafeteria-style, feeding experiments with multiple species, indicated that the best potential for feeding repellents for beavers are extracts from leaves with long lifespans and growing on trees with slow growth rates. Bark extracts have potential as repellents, but specific species which are avoided by beavers have the most potential. No general trends in relation to extracts from bark can be made from these results.

Past research has demonstrated the potential for repellents derived from extracts from non-preferred species for food by North American beavers. Müller-Schwarze *et al.* 1994 demonstrated that the avoidance of red maple (*Acer rubrum*) by beavers in their study sites was in-part due to an extractable chemical in the bark of red maple, and that an extract from the bark of red maple deterred feeding by beavers when samples were available which were painted with extracts from preferred species. Basey *et al.* 1990 showed that North American beavers avoided quaking aspen painted with an extract from the bark of young, juvenile-form quaking aspen resprouting from aspen which had been previously cut by beavers.

Other possible natural repellents to feeding and/or colonization by beavers are pesent. Engelhart and Müller-Schwarze (1995) demonstrated that a variety of predator odors applied to samples reduced feeding by North American beavers in relation to control samples. In addition, artificially scent-marking an area with beaver scent has been shown to prevent colonization by North American beavers (Müller-Schwarze and Heckman, 1980).

The results of this study open the door for a variety of further investigations. How well does the resource-availability hypothesis versus the leaf-longevity hypothesis predict beaver food preferences for other tree species, especially trees with greater variability in leaf lifespans than used in this study? Of utmost importance, are field tests examining whether beavers will change their selection pattern to a less preferred species when the preferred species is painted with a deterrent extract. In addition, can a natural beaver repellent deter beavers from colonizing an unused site? Finally, how long will a natural repellent be effective against beavers?

ACKNOWLEDGMENTS

I thank S. Jenkins for help with numerous aspects of the study and manuscript. I also thank W. Longland, R. Mead, G. Miller, and G. Vinyard for various comments. Funding for this research was provided by a graduate research fellowship from The George Whittell Forest and Wildlife Area, and from the city of Reno, Nevada, USA.

REFERENCES

Aldous, S. E., 1938, Beaver food utilization studies, *J. Wildl. Manage.* **48:**215–222.

Basey, J. M., and Jenkins, S. H., 1993, Production of chemical defenses in relation to plant growth rate, *Oikos* **68:**323–328

Basey, J. M., and Jenkins, S. H., 1995, Influences of predation risk and energy maximization on food selection by beavers (*Castor canadensis*), *Can. J. Zool.* **73:**2197–2208.

Basey, J. M., Jenkins, S. H., and Busher, P. E., 1988, Optimal central-place foraging by beavers: Tree size selection in relation to defensive chemicals of quaking aspen, *Oecologia* **76:**278–282.

Basey, J. M., Jenkins, S. H., and Miller, G. C., 1990, Food selection by beavers in relation to inducible defenses of quaking aspen, *Oikos* **59:**57–62.

Belovsky, G. E., 1984, Summer diet optimization by beaver, *Am. Nat.* **111:**209–222.

Coley, P. D., and Barone, J. A., 1996, Herbivory and plant defenses in tropical forests, *Ann. Rev. Ecol. Sys*, **27:**305–335.

Coley, P. D., Bryant, J. P., and Chapin, F. S. III, 1985, Resource availability and plant antiherbivore defense, *Science* **230:**895–899.

Denslow, J. S., Schultz, J. C., and Vitousek, P. M., 1990, Growth responses of tropical shrubs to treefall gap environments, *Ecology* **71:**165–179.

Denslow, J. S., Vitousek, P. M., and Schultz, J. C., 1987, Bioassays of nutrient limitation in a tropical rain forest soil, *Oecologia* **74:**370–376.

Doucet, C. M., and Fryxell, J. M., 1993, The effect of nutritional quality on forage preference by beavers, *Oikos* **67:** 201–208.

Engelhart, A., and Müller-Schwarze, D., 1995, Responses of beaver (*Castor canadensis* KUHL) to predator chemicals, *J. Chem. Ecol.* **21**(9)**:**1349–1364.

Fowells, J. H., 1965, *Silvics of forest trees of the United States*, U. S. Department of Agriculture, handbook 271.

Fryxell, J. M., and Doucet , C. M., 1991, Provisioning time and central place foraging in beavers, *Can. J. Zool.* **69:**1308–1313.

Fryxell, J. M., Vamosi, S. M., Walton , R. A., and Doucet , C. M., 1994, Retention time and the functional response of beavers, *Oikos* **71:**207–214.

Gulmon, L. L., and Mooney, H. A., 1986. Costs of defense on plant productivity, in: *On the economy of plant form and function*, (Givnish T. J., ed) pp. 681–698, Cambridge University Press, Cambridge.

Jenkins, S. H., 1975, Food selection by beavers: A multidimensional contingency table analysis, *Oecologia* **21:**157–173.

Jenkins, S. H., 1980, A size-distance relation in food selection by beavers, *Ecology* **61:**740–746.

Johnston, C. A., and Naiman, R. J., 1990, Browse selection by beaver: Effects on forest composition, *Can. J. For. Res.* **20:**1036–1043.

Jones, C. E., 1984, Animal Feed, in: *Official methods of analysis of the association of official analytical chemists,* (Williams, S., ed.), pp. 153–169.

Martinsen, G. D., Driebe, E. M., and Whitham, T. G., 1998, Indirect interactions mediated by changing plant chemistry: Beaver browsing benefits beetles, *Ecology* **79:**192–200.

McCanny, S. J., Keddy, P. A., Arnason, T. J., Gaudet, C. L., Moore, D. R. J., and Shipley, B., 1990, Fertility and the food quality of wetland plants: A test of the resource availability hypothesis. *Oikos* **59:**373–381.

McGinley M. A., and Whitham, T. G., 1985, Central place foraging by beavers (*Castor canadensis*): A test of foraging predictions and the impact of selective feeding on the growth form of cottonwoods (*Populus fremontii*). *Oecologia* **66:**558–562.

McKey D. D., 1979, The distribution of secondary compounds within plants. in: *Herbivores: Their interactions with secondary plant metabolites*, (Rosenthal, G. A., Janzen, D. H., eds.), pp. 55–133, New York: Academic.

Müller-Schwarze, D., and Heckman, S., 1980, The social role of scent marking in beaver (*Castor canadensis*), *J. Chem. Ecol.* **6:**81–95.

Müller-Schwarze, D., Shulte, B. A., Sun, L., Müller-Schwarze, A., and Müller- Schwarze, C., 1994, Red Maple (*Acer rubrum*) inhibits feeding by beaver (*Castor canadensis*), *J. Chem. Ecol.* **20:**2021–2034.

Newberry, D. M., and de Foresta, H., 1985, Herbivory and defense in pioneer gap and understory trees in tropical rain forest in French Guiana, *Biotropica* **17:**238–244.

Nolet, B. A., and Rosell, F., 1998, Comeback of the beaver *Castor fiber*: An overview of old and new conservation problems, *Biol. Conserv.* **83:**165–173.

Nolet, B. A., Hoekstra, A., and Ottenheim, M. M., 1994, Selective foraging on woody species by the beaver *Castor fiber*, and its impact on a riparian willow forest, *Biol. Conserv.* **70:**117–128.

Pinkowski, B., 1983, Foraging behavior of beavers (*Castor canadensis*) in North Dakota, *J. Mammal.* **64:**312–314.

Rhoades, D. F., and Cates, R. G., 1976, Toward a general theory of plant antiherbivore chemistry, *Rec. Adv. Phytochem.* **10:**168–213.

Rice, W. R., 1989, Analyzing tables of statistical tests, *Evolution* **43:**223–225.

Sokal, R. R., and Rohlf, J. J., 1981, *Biometry*, 2nd ed. Freeman, San Franscisco, CA.

Southwood, T. R. E., Brown, V. K., and Reader, P. M., 1986, Leaf palatability, life expectancy and herbivore damage, *Oecologia* **70:**544–548.

Swihart, R. K., and Conover, M. R., 1991, Responses of woodchucks to potential garden crop repellents, *J. Wildl. Manage.* **55:**177–181.

16

LONG-TERM POPULATION DYNAMICS OF THE NORTH AMERICAN BEAVER, *CASTOR CANADENSIS*, ON QUABBIN RESERVATION, MASSACHUSETTS, AND SAGEHEN CREEK, CALIFORNIA

Peter E. Busher[1] and Paul J. Lyons[2]

[1]Division of Science and Mathematics
College of General Studies, Boston University
871 Commonwealth Ave.
Boston, Massachusetts 02215
[2]Massachusetts Watershed Initiative
P. O. Box 628
Belchertown, Massachusetts 01007

ABSTRACT

Long-term research and monitoring activities on the Quabbin Reservation in west-central Massachusetts have provided an opportunity to follow changes in numbers and dynamics of a local unexploited beaver, *Castor canadensis,* population over a 45-year period. Beavers returned to the area in the early 1950s, following an absence of more than 200 years. Since then interpretation of aerial photographs, anecdotal reports from early researchers and watershed managers, and starting in 1968, complete annual censuses of all active beaver sites on the 5,018 ha Prescott Peninsula, have identified distinct periods of population change. The first 15 years were characterized by relatively slow growth. This period was followed by another 15 years of very rapid growth of the population. By the early 1980's growth had slowed, and the population experienced a period of rapid decline finally stabilizing at a level approximately 23% of its peak. Various changes in productivity, colony composition, dispersal and interactions with other wildlife species were associated with these periods of population change. The changes observed in the Quabbin Reservation beaver population are comparable to changes in an unexploited beaver population in California. Research on the California beaver population has documented similar periods of rapid growth and subsequent decline.

Beaver Protection, Managment, and Utilization in Europe and North America, edited by Busher and Dzięciołowski.
Kluwer Academic / Plenum Publishers, New York, 1999.

1. INTRODUCTION

The North American beaver, Castor canadensis, was effectively extirpated from most regions of the United States due to intense hunting and trapping pressure by the mid to late 1800s (Jenkins and Busher, 1979; Larson and Gunson, 1983). During the 1900s management programs designed to protect beavers from trapping, coupled with a decline in the market for fur allowed beavers to re-colonize many parts of their former range. Active reintroduction programs were also instituted in many regions. Larson and Gunson (1983), in a summary of the 1981 status of beavers in North America, list 13 of 50 states as having current populations due to introductions, while 36 of 50 states have current populations that grew from remnant populations. One state, New York, is listed as an unknown status. Schulte and Müller-Schwarze (this volume) summarize the growth of beaver populations in the United States and Canada from 1600 to the present. Using data published by Naiman, Melillo, and Hobbie (1986) the beaver population is estimated to be between 6 and 12 million at the present time. The increase in regional beaver populations, particularly in states with high human densities, has in turn led to an increase in the number of beaver nuisance reports. This is true in all the New England states. With a decline in trapping intensity, and low to non-existent predation pressure, many beaver populations in the United States are increasing in a virtually unexploited condition. It is critical to observe and analyze the demographic trends in these populations if biologically sound and effective management plans are to be developed. Conservation biologists in Europe and the United States, can also use this population information to predict the long-term impact of beaver introductions in areas currently without any beavers, or with small populations.

Population data have historically come from regions where beavers were subjected to trapping and hunting pressure. However, in most parts of the United States, there is at present much less trapping and hunting pressure than existed in the past. This allows many beaver populations to exist under relatively natural (not human influenced) conditions. However, few long-term studies of unexploited populations in the United States have been conducted. Long-term (10 or more years) studies have been conducted in Ohio (Svendsen, 1980, 1989), New York (Müller-Schwarze and Schulte, this volume), California (Busher, 1975, 1980, 1987; Hall, 1960; Taylor, 1970), Massachusetts (Bollinger, 1980; Brooks, 1977; Fleming, 1977; Hodgdon, 1978; Howard, 1982; Lancia, 1979; Lyons, 1979a) and Minnesota (Smith and Peterson, 1988). All of these studies have, to some extent, employed livetrapping, marking and observation of animals in open systems. There was also little if any predation pressure on most of these populations, although there were predators (wolves) in the Minnesota study area. That population is also subjected to water fluctuations caused by human activity. Another long-term study is on Isle Royale National Park, Michigan (Shelton, 1966; Shelton and Peterson, 1983). Isle Royale is an isolated island with a healthy predator (wolf) population and can functionally be considered an exploited system.

This paper documents the population dynamics and demographic changes in an unexploited beaver population in western Massachusetts over a 45 year period. We also compare this population to an unexploited population in California over a similar period.

2. METHODS

2.1. Study Areas

The study was conducted on the Prescott Peninsula, Quabbin Reservation, located in west-central Massachusetts (42° 25' N, 72° 20' W) (Figure 1). The Reservation, which con-

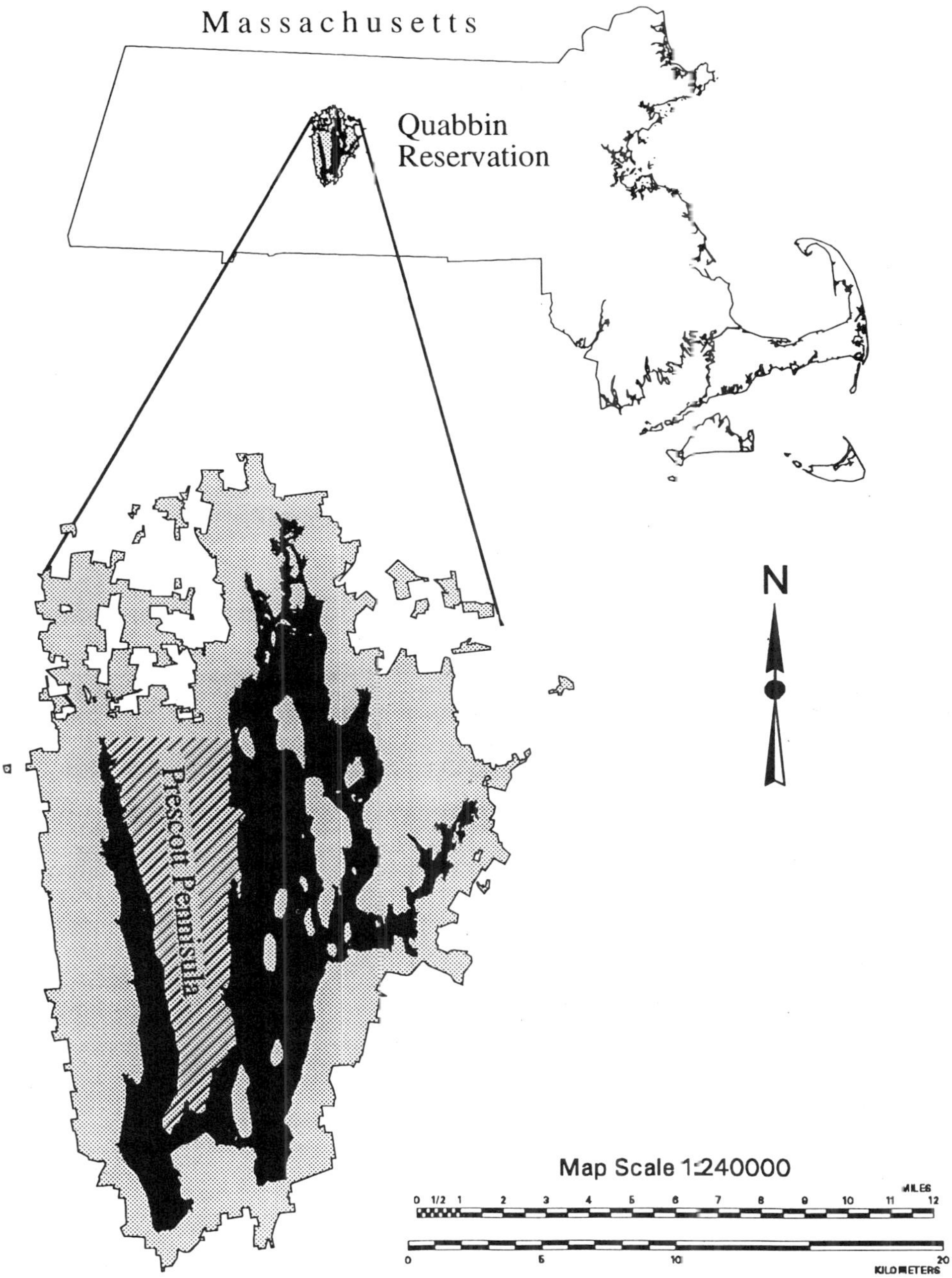

Figure 1. Location of Quabbin Reservation, Massachusetts. The highlighted area shows the location and relative size of the Prescott Peninsula.

tains the watershed and major reservoir for the drinking water supply for the metropolitan Boston area, was created in 1939 by damming the Swift River and Beaver Brook. The area of the reservation is 335 km^2 and the reservoir is 100 km^2. The climate consists of warm, usually moist summers and cold winters with major periods of snow. The construction of the reservoir and creation of the watershed area caused the relocation of 2500 people and the complete destruction of three towns (Hodgdon, 1978; Lyons, 1996). The reservoir was filled to maximum capacity by 1946. The Prescott Peninsula has an area of 50 km^2, is 16 km long and narrows from north to south (4.8 km to 0.6 km). The peninsula is heavily forested (approximately 92%) with dominant deciduous trees being red maple (*Acer rubrum*), oak (*Quercus* spp.), ash (*Fraxinus* spp.), and birch (*Betula* spp.). Eastern white pine (*Pinus strobus*), red pine (*Pinus resinosa*), and hemlock (*Tsuga canadensis*) are the most common conifers on the peninsula (Hodgdon, 1978).

The Quabbin Reservation is managed as a watershed by the Metropolitan District Commission (MDC) an agency of the Commonwealth of Massachusetts. In addition to serving as the water supply for approximately 3 million people in eastern Massachusetts, the reservation provides forest products, recreational opportunities, wildlife observation and research, and cultural resource protection. One of the unique features of the watershed and reservoir is that human use has been strictly controlled since its inception. Until recently, there has been a complete prohibition of trapping and hunting on the reservation and, especially on the Prescott Peninsula, wildlife populations have been allowed to exist with little human interference. Beavers on the Prescott Peninsula have been, and are found on small streams, larger streams, ponds and along the shore of the reservoir.

Sagehen Creek is located on the east slope of the Sierra Nevada in Nevada county, California (Figure 2). The stream originates from natural springs at an elevation of 2400 m and flows 13.5 km to the east where it empties into Stampede reservoir at 1800 m in elevation. The Sagehen Creek basin supports a variety of habitats from moist stream-side meadows and hanging bogs to dry, rocky meadows. The meadows are interspersed with mixed coniferous forests composed of Jeffrey pine (*Pinus jeffreyi*) and lodgepole pine (*Pinus contorta*) at lower elevations and red fir (*Abies magnifica*) and western white pine (*Pinus monticola*) at the higher elevations. Savage (1973) and Trowbridge (1975) describe the vegetation in the basin. Along the stream the dominant shrubs are four willow species (*Salix* spp.) and mountain alder (*Alnus tenuifolia*). Stands of quaking aspen (*Populus tremuloides*) are also found along the stream. The climate is characterized by warm, dry summers and cool wet winters with heavy snowfall. Sagehen Creek is a relatively small stream (width ranges from 2.0 m to 10.0 m) and ponds formed by beaver activity are much smaller than those found on the Prescott Peninsula. Taylor (1970) and Busher (1975) provide detailed maps of sections of the stream. The University of California operates a field station on the stream and the rest of the land adjacent to the stream is controlled by United States Forest Service (Tahoe National Forest).

2.2. Census Methods

Data on the Quabbin beaver population were collected using aerial photographs and anecdotal records from game managers during the early years (1950s and 1960s), livetrapping, mark and release during the 1960s and 1970s, and autumn food cache/activity surveys from 1969 to the present. From 1969 to 1981 the autumn surveys were in conjunction with graduate research projects at the University of Massachusetts, Amherst. From 1981 to the present the surveys have been conducted through the cooperative efforts of the MDC, the United States Fish and Wildlife Service, The Massachusetts Cooperative Wildlife Re-

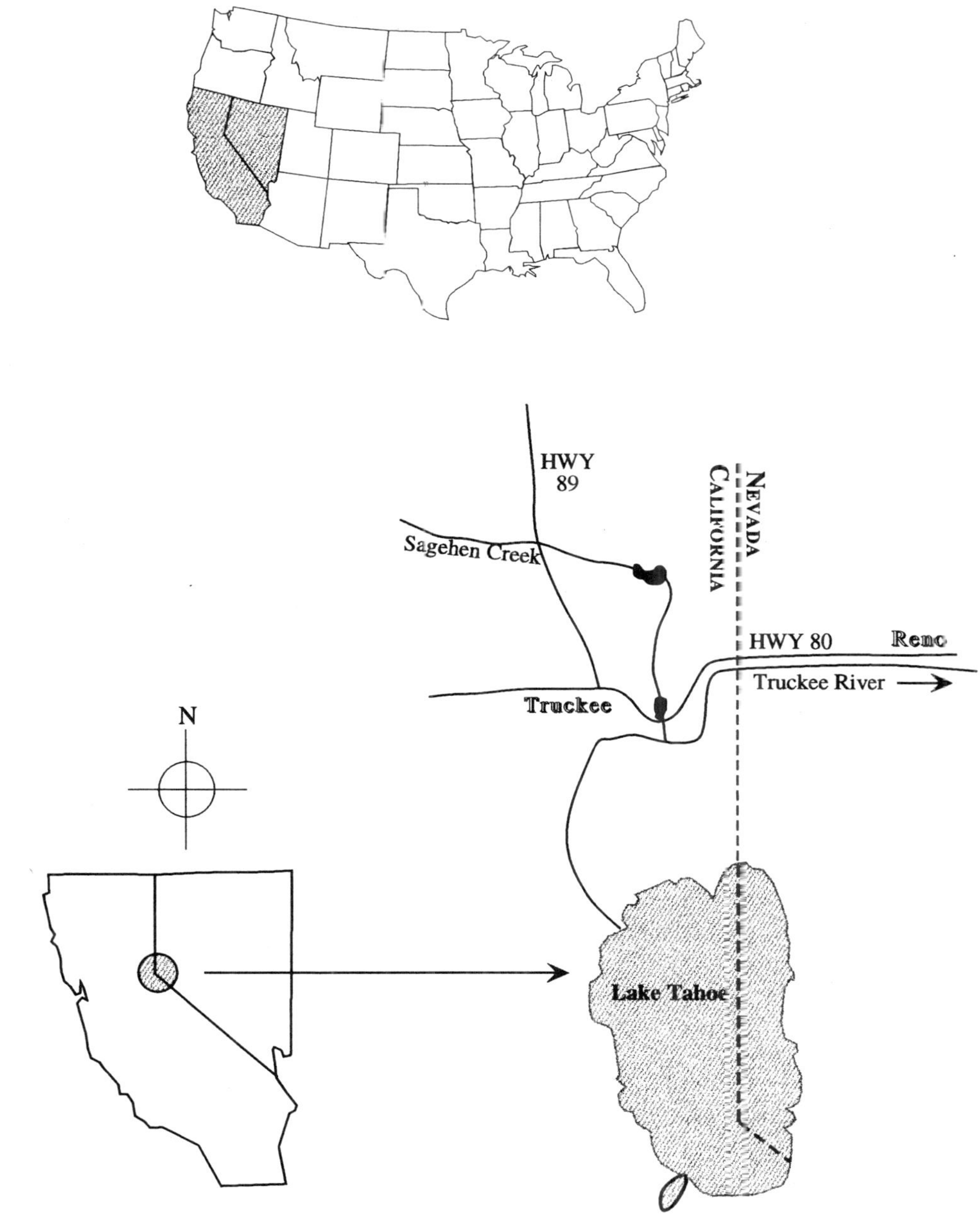

Figure 2. Geographic location of the Sagehen Creek Study area. The location is illustrated by showing its placement in the United States (top map), its placement in the California/Nevada region (map on left), and in a site specific map. In the site specific map Sagehen Creek is shown flowing into a connected reservoir system which is connected to the Truckee River.

search Unit, and the University of Massachusetts, Amherst. The surveys which are conducted in late November or early December each year, consist of having observers walk predetermined routes along streams, ponds, and through other potential beaver habitat on the Prescott Peninsula. A separate boat survey is used to locate active beaver sites along the reservation shoreline. Active beaver sites are identified by the presence of a food cache (usually located near a lodge), recent tree and shrub cutting, and evidence of lodge and dam maintenance (fresh mud and/or freshly cut branches). Observers record the size and composition of the cache and record any beaver sightings. Universal Topographic Map (UTM) coordinates are recorded for each active site.

The Sagehen Creek population has been censused by livetrapping, mark and release methods, coupled with intense observation of the colonies primarily during the summer and autumn. Data reported in this paper end in 1991 although anecdotal information from summer surveys the past few years is also available. Busher (1975, 1980, 1987), Busher, Warner, and Jenkins (1983), and Taylor (1970) provide more detailed information on the actual livetrapping technique.

2.3. Initial History of the Two Populations

Shaw (1948) reported that by the early 1900s beavers were returning to Massachusetts after being extirpated 200 years earlier. Beavers were first observed on the Prescott Peninsula in 1952 (Hodgdon, 1978) a few years after the reservoir had been filled. This paper presents the reconstruction of the Quabbin beaver population on the Prescott Peninsula from that time to the present (1997).

The Sagehen Creek population was started by the introduction of four adults (two pairs) by the California Department of Fish and Game in 1945 (Hensley, 1946). The population was initially studied by Joseph Hall (1952–1955). He was followed by Richard Gard (1954–1957), David Taylor (1958–1970), and Peter Busher (1973 -1980). Ellen Woodard studied the population from 1984–1991.

Both populations have received extensive attention from researchers examining various aspects of the populations. The Quabbin population was initially studied in the mid 1960s by Stanton (1965) and Keiper (1966) who investigated waterfowl use and macroscopic bottom fauna associated with beaver areas respectively. Hodgdon (1971, 1978) provided an extensive analysis of the population dynamics and Lancia (1979) examined the activity period of beavers using radio telemetry. Later, Brooks (1977) and Fleming (1977) tested the effects of sterilization on family structure (Brooks, Fleming, and Kennelly, 1980). This work was followed up by Bollinger (1980) and Lyons (1979a, 1979b). Olsen (1980) and Howard (1982) also contributed by examining beaver physiological condition and habitat suitability respectively. Busher (1991, 1996) investigated the selection and use of woody species during the autumn food caching period and Busher and Lyons (1988) conducted the last extensive livetrapping, mark and release effort on the Prescott Peninsula.

On Sagehen Creek, Joseph Hall investigated the use of aspen and willow by beavers (Hall, 1956, 1960). Richard Gard studied the impact of beaver activity on the aquatic macrofauna (Gard, 1958, 1961), and David Taylor conducted a long-term analysis of population and habitat changes (Taylor, 1970). In the early 1970s, Peter Busher constructed the first radio transmitters used on beavers in the United States and examined beaver movements and activities (Busher 1975). Busher followed this study by examining the population dynamics and behavior of beavers on Sagehen Creek (Busher, 1980, 1987; Busher and Jenkins, 1985; Busher, *et al.*, 1983). From 1984 to 1991 Ellen Woodard examined the behavior, activity patterns and foraging behavior of the Sagehen Creek population (Woodard, 1994).

3. RESULTS

3.1. Prescott Peninsula Population

The Prescott Peninsula population history (Figure 3) is typified by five phases. The initial phase (1952–1968) was a time of very slow population growth. During this time period the number of colonies increased from 2 to 16, which is a change of 0.82 colonies per year.

A second phase was observed beginning in 1968 (when extensive annual ground surveys began) and continued through 1975. During this time period the number of colonies increased from 16 to 46, which is an average change of 4.29 colonies per year. An increase in the number of shore colonies contributed to this rapid increase phase (Table 1).

The third phase in the population occurred between 1975 and 1983, which was a time of fluctuating population levels at a high density. Overall, the population declined by 4% (46 to 44 colonies) during this period, although year-to-year fluctuations were as high as 20%. During the second and third phases (1969–1983) there was rapid colonization of new sites on both the peninsula and the shore, and by 1983 many marginal sites were occupied.

A fourth phase occurred between 1983 and 1988 when the number of colonies declined from 44 to 12. This is a mean annual decline of 6.4 colonies. The decline during this period was precipitated by a reduction in the number of shore colonies from 13 in 1983 to one in 1985, 1986 and 1987. By 1988 no shore colonies were present.

The period from 1988 to the most recent survey in 1996 represents the fifth phase in the population. During this time the number of colonies has ranged from a low of 10 to a high of 15. The average change (a decrease) is 0.125 colonies per year during this period of relative stability at low population levels.

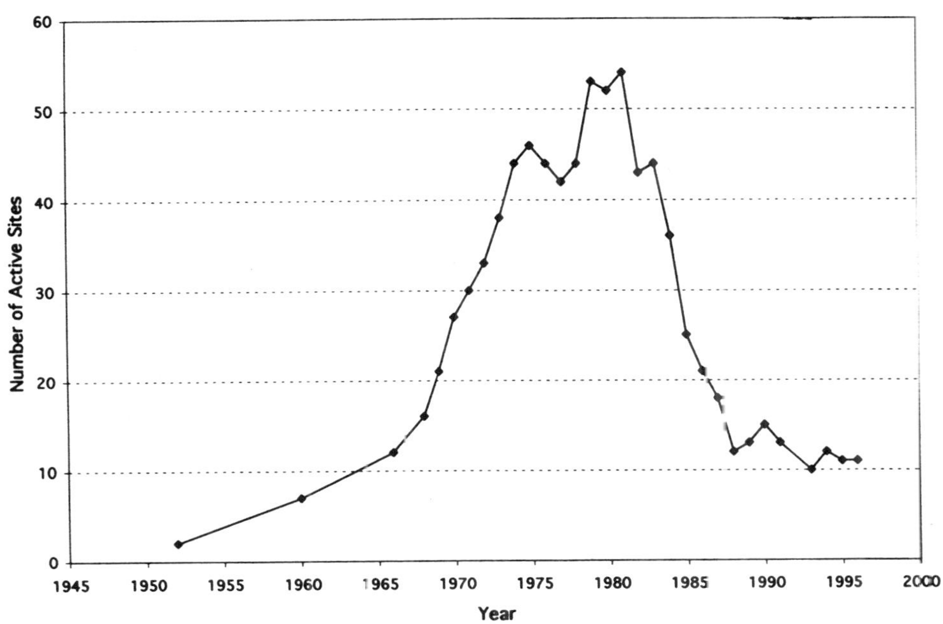

Figure 3. Results of beaver population surveys on Prescott Peninsula, Quabbin Reservation, Massachusetts, 1952–1996.

Table 1. Number of active beaver colonies on Prescott Peninsula, Quabbin Reservation, Massachusetts, 1952–1996

	Number of active sites		
Year	Interior	Shore	Total
1952	2	0	2
1960	7	0	7
1966	12	0	12
1968	16	0	16
1969	18	3	21
1970	21	6	27
1971	24	6	30
1972	22	11	33
1973	23	15	38
1974	26	18	44
1975	—	—	46
1976	30	14	44
1977	28	14	42
1978	30	14	44
1979	36	17	53
1980	35	17	52
1981	39	15	54
1982	29	14	43
1983	31	13	44
1984	24	12	36
1985	24	1	25
1986	20	1	21
1987	17	1	18
1988	12	0	12
1989	13	0	13
1990	13	2	15
1991	11	2	13
1993	7	3	10
1994	9	3	12
1995	11	0	11
1996	10	1	11

The age-class composition of the population (Table 2) is drawn from data presented by Hodgdon (1978) and from Busher and Lyons (1988). In 1969 at the beginning of the increase phase there were 35.6% adults, 5.8% 2-year-olds, 25% yearlings, and 33.7% kits. By 1974 when the population was approaching the first peak Hodgdon (1978) reported a total population size of 257 animals. The composition in 1974 was 37.4% adults, 3.5% 2-year-olds, 28.8% yearlings, and 30.4% kits. The total number of animals, and the age composition changed dramatically by 1988. Busher and Lyons (1988) reported only 40 animals in the population and an age-class composition with 47.5% adults, 20% 2-year-olds, 17.5% yearlings, and 15% kits. This year (1988) represents the end of the decline phase and the beginning of the stabilization phase at low population levels.

3.2. Sagehen Creek Population

Sagehen Creek includes a much smaller area than the Prescott Peninsula and supported a smaller number of beavers and colonies. For Sagehen Creek the number of bea-

Table 2. Age-class composition on the Prescott Peninsula, Massachusetts, during selected years

Year	Adult	2-year-old	Yearling	Kit	Total number
1969	35.6	5.8	25	33.7	104
1974	37.4	3.5	28.8	30.4	257
1988	47.5	20	17.5	15	40

In 1969 the population was at the beginning of the increase phase in 1974 the population was at the end of the increase phase and entering the high density phase, and in 1988 the population was entering the low density phase.

vers (not the number of colonies) is represented (Figure 4). Additionally, the stream is bisected by California State Highway 89, and the section west of the highway has been studied most intensively and only data from this section are presented here. The Sagehen Creek population, much like the Prescott Peninsula population, had an initial phase of slow growth, a second phase of more rapid increase, followed by a third phase of relative stability at high numbers, and a fourth phase of sharp decline. The decline was followed by only a few years of relative stability at low numbers. The Sagehen Creek population, unlike the Prescott Peninsula population, had a fifth phase characterized by rapid population growth. This phase was in turn followed by a second decline phase (1979–1991).

The initial period of slow growth was between 1948–1956. Beavers colonized the eastern sections of the stream first, but by the late 1940s and early 1950s were also using areas west of the highway. The population increased from an estimate of two animals in 1948 to 10 in 1956. The average increase was 1.0 beavers per year during this period.

During the second phase, from 1956 to 1959, there was a sharp increase from 10 animals to approximately 20 animals for an average annual increase of 3.3 animals. The

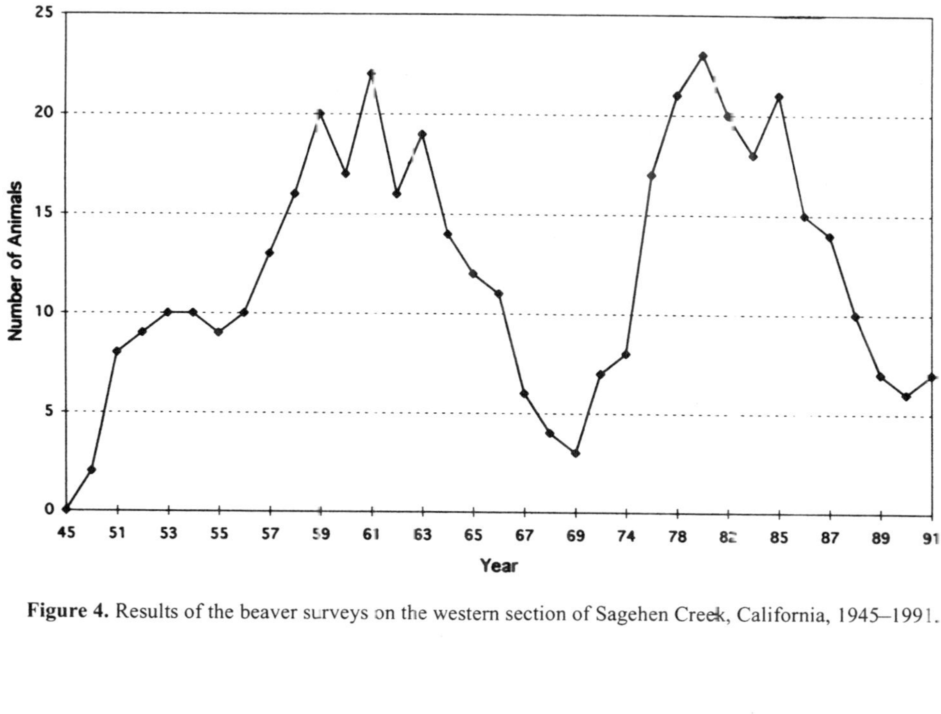

Figure 4. Results of the beaver surveys on the western section of Sagehen Creek, California, 1945–1991.

beavers colonized the upper sections of the stream during this time and the family groups became well established.

The period 1959–1963 (phase three) represented 5 years of relative stability in number of animals (ranging from 18–22). Twenty animals were present in 1959 and 20 were present in 1963. Taylor (1970) reports four well established family groups on this section of stream during this period.

The fourth phase (1963–1969) was characterized by a sharp decline in the number of beavers. Taylor (1970), reported that by 1969 there was only one active colony on this section of stream and it was composed of only one pair of adults. There was an average decrease in the number of beavers during this period of 3.0 animals per year.

During the fifth phase (1969–1979) the number of beavers increased from 2 to 23, which is an average increase of 2.1 beavers per year (Busher, 1987). The number of colonies ranged between two and four during this phase. In 1982 R. N. Brown conducted an unpublished, one summer study and reported a small decline in number to 20 beavers. This was the beginning of a second decline phase.

Ellen Woodard studied the population from 1984 to 1991, and documented the decrease in number of beavers during these years. Woodard reported that the population declined from 18 animals in 1984 to 7 animals in 1991. The beaver numbers were affected by two drought years, and animals moved from the western section of the stream to the eastern section at that time (Woodard, 1994; and pers. comm.).

4. DISCUSSION

The Prescott Peninsula and Sagehen Creek beaver populations show a similar pattern of growth, stabilization and decline. The populations are different in that the Prescott beavers have stabilized at a low density while the Sagehen beavers increased a second time and then declined again. While the dynamics of the two populations are similar, the areas are quite different in geographic location, size, vegetation, topography, elevation, and stream characteristics. The Prescott Peninsula has a number of large persistent ponds fed by small streams, while Sagehen Creek is one small stream (average width is approximately 5 m). From both management and conservation biology points of view the growth, peak and subsequent decline of both populations provide insight into beaver population dynamics in unexploited populations.

The Prescott Peninsula population initially exhibits the classic sigmoid growth pattern that is often associated with populations colonizing new habitat (Caughley, 1977, Elton, 1958). The rapid decline between 1984 and 1988 indicates that beavers had colonized all available habitats including many marginal areas. Hall (1960) suggested that beavers will utilize a food resource in an area then move to another area leaving the first site vacant for a number of years. This behavior of habitat rotation allows food resources to rebound after beaver occupation. On the Prescott Peninsula many marginal areas were occupied by the early 1980s. For beavers living in these marginal areas, once the food resources were depleted and they were forced to move, there were few if any high quality unoccupied sites available on the peninsula. Thus, beavers were forced to leave the peninsula or perish. The drastic decline in the number of shore colonies from 1988 to the present suggests that shore sites are not optimal for beaver occupation. The shore sites are subject to water fluctuations during drought years when the reservoir is not full. However, the reservoir has not been at low levels every year since 1988.

Further evidence that the Prescott Peninsula is at carrying capacity for beavers is the change in the age-class composition that occurred between the time of peak population size (early 1970s) and 1988 (the last trap, mark and recapture study). In 1974 Hodgdon reported that 59.2% of the population was pre-reproductive animals (yearlings and kits) (Hodgdon, 1978). In 1988 only 32.5% of the population was pre-reproductive. In 1974, 2-year-old beavers were only 3.5% of the total population while in 1988 they were 20%. This suggests a reduction in dispersal of these young adult animals. Lack of dispersal in beavers is observed at very high population densities associated with habitat saturation (Bergerud and Miller, 1977; Novakowski, 1965; Busher, 1987; Müller-Schwarze and Schulte, this volume). However, lack of dispersal can also occur at lower densities or in areas where little or no quality habitat is available (after all the marginal sites have been used and abandoned). When Hodgdon (1978) observed a high rate of 2-year-old dispersal at high densities there were still suitable (although marginal) areas available for dispersing animals. However, once these marginal sites had been occupied and then later abandoned, they were not suitable sites able to sustain dispersing 2-year-old beavers.

A second factor that may have influenced the beaver population is the occurrence of whitetail deer, *Odocoileus virginianus*, on the peninsula. The deer population was very dense through the 1980s. Deer reduce the regeneration of trees by eating the saplings. There is evidence that deer activity on the peninsula was inhibiting forest regeneration. Thus, woody species that beaver would have used for food after they regenerated were being removed by deer. This interaction between deer, regenerating forage trees, and beaver could account for the most recent population phase (a low density, stable population). A controlled deer hunt on the peninsula was instituted in 1990 and an increase in forest regeneration is being observed. The Prescott Peninsula beaver population may show a similar rebound phase much like the Sagehen Creek population in the future.

Both study areas have been subject to beaver habitat classification studies. Howard and Larson (1985) developed a stream habitat classification system for beaver using information from the Prescott Peninsula population. They found that water reliability variables significantly affect colony site longevity, while food availability was considered less important. However, they do note that the relationship between water reliability and aquatic plant productivity is critical since beaver may use aquatic plants for food during both summer and winter.

Beier and Barrett (1987) evaluated habitat use and impact by beavers in the area around Sagehen Creek. They state that, "Stream gradient, stream depth, and stream width were clearly the most important factors related to beaver habitat use..." They continue by saying that, "Vegetation variables added little explanatory power to the functions." They considered the suitable beaver habitat in the entire Truckee River basin (including Sagehen Creek) to be saturated with beavers when the study was conducted (May–August, 1985). The second decline phase of the Sagehen Creek beaver population in the late 1980s supports this view of habitat saturation. It is possible that after two periods of high beaver numbers (early 1960s, 1970s) the vegetation in potentially marginal habitat may have been effectively depleted. Stream width is a critical variable in the Beier and Barrett study. Many sections of the western part of Sagehen Creek are 5.0 m or less in width well below the average width of 8.1 m associated with active colonies. Thus, the combination of marginal physical requirements of the stream, coupled with intensive occupation and food utilization by beavers may have led to the latest population decline. Summer and autumn surveys on Sagehen Creek are being planned for 1999 to determine the extent of re-colonization of the area.

These two unexploited populations, while geographically different, do provide a picture of what can be expected when beavers are introduced to or move into an area. Habitat classification studies such as those by Beier and Barrett (1987), Hartman (1996), Howard and Larson (1985), and Slough and Sadlier (1977), are useful in predicting areas that will support long-term beaver occupation. However, as is evident from the Prescott Peninsula population, even marginal sites will be utilized in virgin habitat when populations are dense. These data from two unexploited populations will aid conservation biologists and wildlife managers when they develop long-term (10–50 year) beaver management plans.

ACKNOWLEDGMENTS

Long-term studies require the cooperation of many people. The authors are grateful to all of the researchers who have studied the beaver populations on Quabbin Reservation, and Sagehen Creek. We especially acknowledge those individuals who have worked with us in the field, and all whose own research has stimulated our thinking on the dynamics of beaver populations. Many people helped conduct the annual beaver surveys on Quabbin Reservation over the years. While it is impossible to identify all of them here, the authors would like to acknowledge the following people who played a key role in organizing those surveys: Harry Hodgdon, Joseph Larson, Wendell Dodge, Jim Kennelly, Jack Finn, Becky Field, Richard Lancia, Rob Brooks, Mike Fleming, Karen Bollinger, Carolyn Boardman, Rebecca Howard, Suzie Von Oettingen, John Organ and Glenn Olsen. The Metropolitan District Commission kindly provided access to Quabbin Reservation for the surveys, and provided logistical support.

I (PEB) am especially indebted to Joseph Hall and David Taylor who shared with me their knowledge of the Sagehen Creek area, and beaver ecology. Stephen Jenkins has also provided insight into beaver population dynamics. Collection of the Sagehen Creek Data between 1973–1980 was made possible by support from San Francisco State University and the University of Nevada, Reno. Quabbin Reservation research conducted during 1988 was supported by a grant to the authors from The Center for Field Research–Earthwatch. We are grateful for the field support provided by the Earthwatch volunteers during that time period.

REFERENCES

Beier, P., and Barrett, R. H., 1987, Beaver habitat use and impact in the Truckee River basin, California, *J. Wildl. Manage.* **51:**794–799.

Bergerud, A. T., and Miller, D. R., 1977, Population dynamics of Newfoundland beaver, *Can. J. Zool.* **55:**1480–92.

Bollinger, K. S., 1980, Scent-marking behavior of beaver (*Castor canadensis*), M.S. thesis, Univ., Massachusetts, Amherst, p. 186.

Brooks, R. P., 1977, Induced sterility of the adult male beaver (*Castor canadensis*) and colony fecundity, M.S. thesis, Univ., Massachusetts, Amherst, p. 90.

Brooks, R. P., Fleming, M. W., and Kennelly, J. J., 1980, Beaver colony response to fertility control: evaluating a concept, *J. Wildl. Manage.* **44:**568–575.

Busher, P. E., 1975, Movements and activities of beavers, *Castor canadensis*, on Sagehen Creek, California, M.A. thesis, San Francisco State univ., San Francisco, p. 86.

Busher, P. E., 1980, The population dynamics and behavior of beavers in the Sierra Nevada, Ph.D. dissert., Univ. Nevada, Reno, p. 147.

Busher, P. E., 1987, Population parameters and family composition of beaver in California, *J. Mamm.* **68:**860–864.

Busher, P. E., 1991, Food caching behavior by the North American beaver, *Castor canadensis*, in western Massachusetts, in: *Global Trends in Wildlife Management*, (B. Bobek, K. Perzanowski, and W. Regelin eds.), pp. 111–114, Transactions of the 18th IUGB Congress, Swiat Press, Krakow-Warszawa, Poland.

Busher, P. E., 1996, Food caching behavior of beavers (Castor canadensis): Selection and use of woody species, *Am. Midl. Nat.* **135:**343–348.

Busher, P. E., and Jenkins, S. H., 1985, Behavioral patterns of a beaver family in California, *Biol. Behav.* **10:**41–54.

Busher, P. E., and Lyons, P. J., 1988, Quabbin beaver project: Research summary, Unpubl. research report for The Center for Field Research-Earthwatch (October 1988), p. 1–8.

Busher, P. E., Warner, R. J., and Jenkins, S. H., 1983, Population density, colony composition, and local movements in two Sierra Nevadan beaver populations, *J. Mamm.* **64:**314–318.

Caughley, G., 1977, *Analysis of Vertebrate Populations*, Wiley, New York and London.

Elton, C. S., 1958, *The Ecology of Invasions by Animals and Plants*, Methuen & Co. LTD., London.

Fleming, M. W., 1977, Induced sterility of the adult female beaver (*Castor canadensis*) and colony fecundity, M.S. thesis, Univ., Massachusetts, Amherst, p. 59.

Gard, R., 1958, The effects of beaver on trout in Sagehen Creek, California, Ph.D. dissert., Univ. California, Berkeley, p. 110.

Gard, R., 1961, The effects of beaver on trout in Sagehen Creek, California, *J. Wildl. Manage.* **25:**221–242.

Hall, J. G., 1956, Willow and aspen in the ecology of beaver on Sagehen Creek, California, Ph.D. dissert., Univ. California, Berkeley, p. 88.

Hall, J. G., 1960, Willow and aspen in the ecology of beaver on Sagehen Creek, California, *Ecology* **41:**484–494.

Hartman, G., 1996, Habitat selection by a European beaver (*Castor fiber*) colonizing a boreal landscape, *J. Zool. Lond.* **240:**317–325.

Hensley, A. L., 1946, A progress report on beaver management in California, *Calif. Fish and Game* **32:**87–99.

Hodgdon, H. E., 1971, Some sexual differences within a marked beaver colony, M.S. thesis, Univ., Massachusetts, Amherst, Hodgdon, H. E., 1978, Social dynamics and behavior within an unexploited beaver (*Castor canadensis*) population, PhD dissert, Univ. Massachusetts, Amherst, p. 292.

Howard, R. J., 1982, Beaver habitat classification in Massachusetts, M.S. thesis, Univ. Massachusetts, Amherst, p. 67.

Howard, R. J., and Larson, J. S., 1985, A stream habitat classification system for beaver, *J. Wildl. Manage.* **49:**19–25.

Jenkins, S. H., and Busher, P. E., 1979, *Castor canadensis*, *Mamm. species* **120:**1–8.

Keiper, R., 1966, The distribution and faunal succession of the macroscopic bottom fauna in three different aged beaver ponds, M.S. thesis, Univ., Massachusetts, Amherst, p. 96.

Lancia, R. A. 1979, Year-long activity patterns of radio-marked beaver (*Castor canadensis*) , Ph.D. dissert., Univ. Massachusetts, Amherst, p. 146.

Larson, J. S., and Gunson, J. R., 1983, Status of the beaver in North America, *Acta Zool. Fenn.* **174:**91–93.

Lyons, P. J., 1979a, Effects of induced sterility on reproduction and dispersal patterns in beaver colonies, M.S. thesis, Univ. Massachusetts, Amherst, p. 123.

Lyons, P. J., 1979b, Productivity and population structure of western Massachusetts beavers, *Trans. Northeast Sect. Wildl. Soc. Fish Wildl. Conf.* **36:**176–187.

Lyons, P. J., 1996, Beaver research and management on a public water supply, in: *Living with Wildlife: The Biology and Sociology of Suburban Deer and Beaver (*A. N. Rowan, and J. C. Weer, eds.), pp. 141–151, Tufts Center for Animals and Public Policy, N. Grafton, MA. Naiman, R. J., Melillo, J. M., and Hobbie, J. E., 1986, Ecosystem alteration of boreal forest streams by beaver (*Castor canadensis*), *Ecology* **67:**1254–1269.

Novakowski, N. S., 1965, Population dynamics of a beaver population in northern latitudes, Ph.D. dissert., Univ. Saskatchewan, Saskatoon, p. 154.

Olsen, G., 1980, Beaver (*Castor canadensis* Kuhl) temperature monitored by radiotelemetry, M.S. thesis, Univ., Massachusetts, Amherst, p. 86.

Savage, W., 1973, Annotated check list of vascular plants of Sagehen Creek drainage basin, Nevada County, California, *MadroÒo* **22:**115–139.

Shaw, S. P., 1948, The beaver in Massachusetts, Mass. Dept. Conserv., Boston, Res. Bull. No. 11, p. 45.

Shelton, P. C., 1966, Ecological study of beavers, wolves, and moose in Isle Royale National Park, Ph.D. dissert., Purdue Univ., Lafayette, Indiana, p. 308.

Shelton, P., and Peterson, R., 1983, Beaver, wolf, and moose interactions in Isle Royale National Park, USA, *Acta. Zool. Fenn.* **174:**265–266.

Slough, B. G., and Sadleir, R. M. F. S., 1977, A land capability classification system for beaver (*Castor canadensis* Kuhl), *Can. J. Zool.* **55:**1324–1335.

Smith, D. W., and Peterson, R. O., 1988, The effects of regulated lake levels on beaver in Voyagers National Park, Minnesota, National Park Service, Research/Resources Management Report MWR-11, (July 1988), p. 84.

Stanton, P. B., 1965, An evaluation of waterfowl utilization on three age classes of beaver impoundments with emphasis on the black duck (*Anas rubripes*), M.S. thesis, Univ., Massachusetts, Amherst, p. 112.

Svendsen, G. E., 1980, Population parameters and colony composition of beaver (*Castor canadensis*) in southeast Ohio, *Am. Midl. Nat.* **104:**47–56.

Svendsen, G. E., 1989, Pair formation, duration of pair-bonds, and mate replacement in a population of beavers (*Castor canadensis*), *Can. J. Zool.* **67:**336–340.

Taylor, D., 1970, Growth, decline, and equilibrium in a beaver population at Sagehen Creek, California, Ph.D. dissert., Univ. California, Berkeley, p. 169.

Trowbridge, B. A., 1975, Plants of upper Sagehen Creek drainage basin, Nevada County, California, M.A. thesis, San Francisco State University, San Francisco, p. 106.

Woodard, E. L., 1994, Behavior, activity patterns and foraging strategies of beaver (*Castor canadensis*) on Sagehen Creek, Callifornia, Ph.D. dissert., Univ. California, Berkeley, p. 311.

17

BEHAVIORAL AND ECOLOGICAL CHARACTERISTICS OF A "CLIMAX" POPULATION OF BEAVER (*CASTOR CANADENSIS*)

Dietland Müller-Schwarze[1] and Bruce A. Schulte[2]

[1]State University of New York
College of Environmental Science and Forestry
Syracuse, New York 13210
[2]Department of Biology
Providence College
Providence, Rhode Island 02918-000

ABSTRACT

We describe the characteristics of an unexploited (unharvested) beaver population in New York State. The results are based on data collected from 1984 to 1993. The following features differed from exploited (harvested) populations: Beavers colonized steeper stream gradients; Families contained a large percentage of 3-year-olds (young that stay one more year); Preferred tree species became depleted and beaver fed on less palatable species; and they extended their trails to more distant foraging areas. In other ways, our population resembled trapped populations: Densities of colonies per stream kilometer did not differ from those reported in other studies; Body weights did not differ from those in other studies; Family size was within the range of other populations; The numbers of scent mounds in territories are not reduced or increased; and The response to artificial scent marks does not differ in intensity from other areas. In conclusion, the beavers are adaptable and need little management by humans. Beaver management and control of beaver numbers are driven by conflicts with humans, and not by any needs of the beavers themselves.

Beaver Protection, Managment, and Utilization in Europe and North America, edited by Busher and Dzięciołowski.
Kluwer Academic / Plenum Publishers, New York, 1999.

1. INTRODUCTION

1.1. Exploited and Unexploited Beaver Populations

What is the fate of a beaver population when we leave it alone and allow it to grow freely? Here we use the term "climax population" for unexploited (unharvested) beavers in a large area who have occupied all suitable sites at one time or another. However, the beavers may rotate through many sites in a cycle of occupancy and vacancy. Their overall population size may be stable at higher or lower levels at different times.

Specifically, from among the many possible variables, do areas colonized, distances between colonies, inter-colony strife, family size, family composition, body weights, food supply, and foraging distances change measurably as the population increases and reaches saturation? Over evolutionary time, beaver populations have always grown to, and maintained themselves at certain, albeit fluctuating levels. Predation, diseases, and food depletion have imposed limits to extreme population growth. In the modern landscape, beavers occur in reintroduced, small and growing populations, in established, periodically exploited (harvested by trapping) populations, and in unexploited (not trapped), saturated populations. Unexploited populations represent the original state, but now beavers are often in conflict with humans in most areas. Conflicts occur regardless of whether these areas are developed and built-up, used for recreation, used for growing and/or harvesting timber, or used for agriculture. These unharvested populations can teach us what we can expect if the newly introduced beavers in many areas multiply to saturation levels.

In Alaska, Boyce (1981) compared an exploited (harvested) beaver population on the Chena River to an unexploited (unharvested) population on the Birch River. He found that neither the density of the populations, nor the fecundity of females, as measured by numbers of embryos, differed between the two beaver populations. These populations only differed in the relationship of age and body size. In the exploited population, females reproduced at a younger age. Coupled with this, the 5- and 6-year old females were smaller than their counterparts in the unexploited population. Many of these smaller females did not survive until the ages of 12 to 14 years. Because the body weight of younger females was depressed, body size correlated with age at the exploited population, while adults of all ages were of the same (larger) size in the unexploited population (Boyce, 1981).

1.2. Questions Asked

In their natural state beavers have always lived in freely growing, eventually saturated populations. While the current conflicts between beavers and humans are real, we ask here what effect the high population density has on the beavers. Do the beavers suffer in any way from the high population density? Can we detect measurable differences between our unexploited population and other, exploited populations, such as a higher colony density, a depleted food base, lower body weights, smaller family size, or altered family composition?

In this chapter we present data on various aspects of our study population and compare them with those from either our other study areas in New York, or the literature, depending on available data.

2. STUDY AREAS

2.1. Main Study Area—Allegany State Park

Data were collected over a 10 year period (1984–1993) from an unexploited beaver population in Allegany State Park (ASP) located in western New York (Figure 1). Allegany State Park includes an area of about 250 km^2 and contains about 70 km of streams. The park is covered by northern hardwoods, including sugar maple (*Acer saccharum*), red maple (*A. rubrum*), black cherry (*Prunus serotina*), American hornbeam (*Carpinus caroliniana*), American beech (*Fagus grandifolia*), white ash (*Fraxinus americana*), and also willows (*Salix* spp.) and quaking aspen (*Populus tremuloides*). Beavers first appeared in the park in 1937 (Shadle and Austin, 1939). They have not been harvested in ASP since 1974 when 93 beavers were removed, leaving only 14 active sites. From 1983 to 1988 the number of active sites increased from 55 to 64 (ASP Draft Master Plan 1992). During the same time span the rate of occupancy of potential beaver sites increased from 50% to 58%. During the years 1989–93 the number of active sites ranged from 50 to 55 (Schulte, 1993). In spring 1998 the occupancy rate of potential sites was 42% (DM-S own survey).

The beavers at Allegany State Park are popular with visitors, and tours of beaver ponds are among the best-attended offerings at the park. Beaver ponds also offer many visitors recreational opportunities, such as good fishing and bird watching. On the other hand, managers consider the beaver population too dense, and "management" (beaver removal) is often suggested. Flooding of roads and the destruction of trout spawning habitat are the most frequent complaints leveled against the beavers.

2.2. Other Study Areas

Our other three study areas in New York are located in the Adirondack Mountains: 1) Huntington Wildlife Forest (HWF) near Newcomb in the center of the Adirondacks (unexploited); 2) Fulton County, at the southern fringe of the Adirondack ecosystem (exploited); and 3) Cranberry Lake Biological Station (unexploited), in the western Adirondacks (Figure 1).

3. RESULTS

3.1. Stream Gradients

Beavers generally colonize stream sections with gradients of 1% or less. Do beavers colonize steeper areas after the optimal sites are taken? To address this question, we surveyed a sample of six occupied beaver sites, each in three different areas: ASP, HWF, and Fulton County. In each area, we chose the six steepest sites we could find. Figure 2 shows the stream gradients in the three areas. In ASP the two steepest sites had stream gradients of 6.44% and 6.32%, respectively, while at HWF gradients ranged from 0.15% to 2.0%, and in Fulton County from 0.31% to 0.78%. The differences between ASP and each of the other two areas are significant ($P = 0.02$, two-tailed t-test).

3.2. Population Density

At ASP, the density of colonies per stream km ranged from 0.24 to 1.14, depending on watershed and year (Table 1). The overall mean density in 1990 was 0.52 colonies per km stream, 0.58 in 1991, and 0.55 in 1992, while in the years 1990–92 the highest density observed (France Brook) was 0.91 colonies/km, and the lowest (English Stoddart Brook) was 0.20 (Schulte, 1993). Compared with populations in other studies the overall density at ASP is not particularly high. Reported densities range from 0.40 colonies/km in Alaska (Boyce, 1974) to 1.09 in New Brunswick (Nordstrom, 1972), as shown in Table 1. However, locally, at lower Red House Brook, one of the highest densities (1.14 colonies/km)

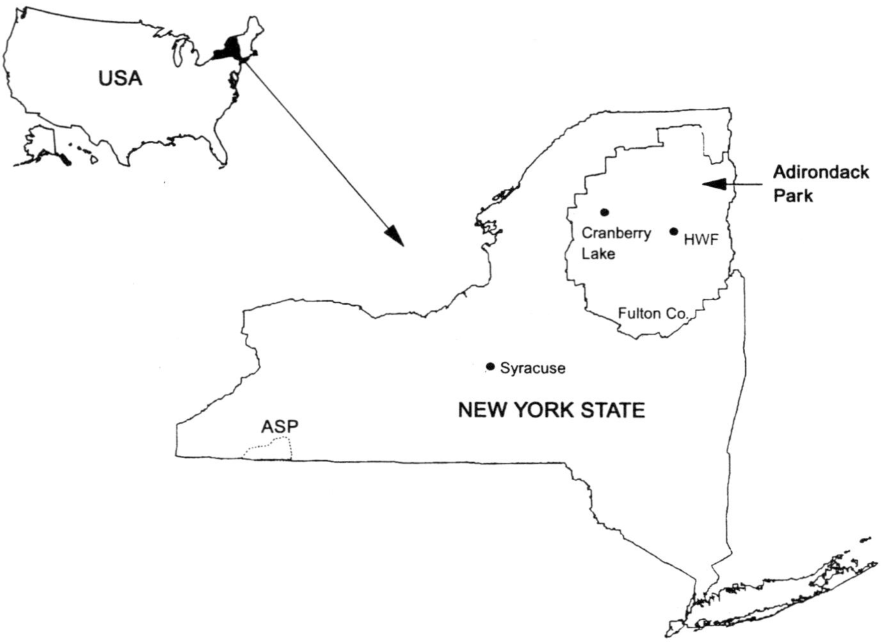

Figure 1. Map of the study areas in New York.

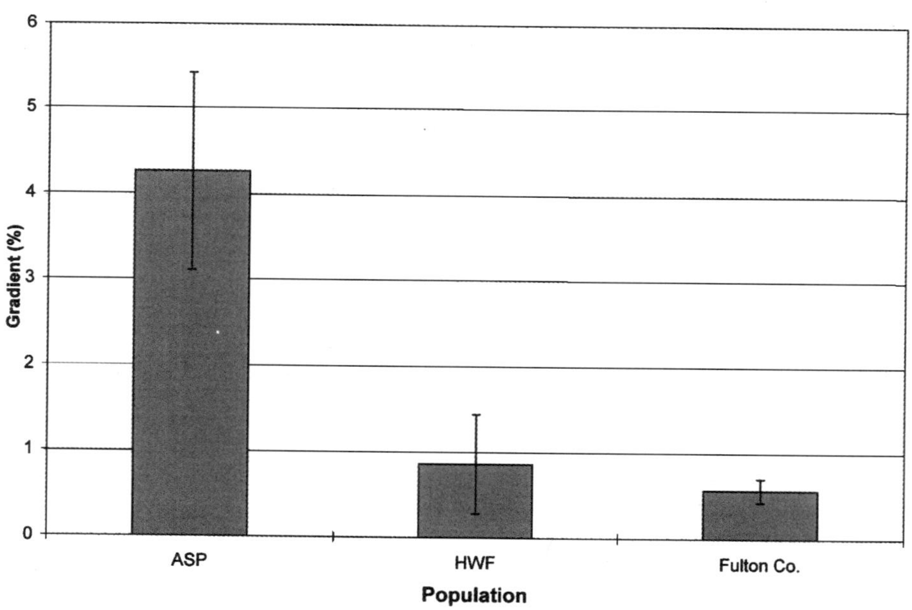

Figure 2. Stream gradients at Allegany State Park (ASP), Huntington Wildlife Forest (HWF), and Fulton County, all in New York. For each area, the 6 steepest gradients were measured.

Table 1. Comparison of densities of beaver populations (colonies per stream kilometer) in different areas

ASP, lowest (Mackintosh Bk.)	0.24	This study
Alaska	0.40	Boyce, 1974
ASP, overall 1990	0.53-0.60	This study
Fulton Co.	0.54	Welsh & Müller-Schwarze
ASP, overall 1991	0.55-0.625	This study
Massachusetts: Quabbin Reservation	0.55	Hodgdon, 1978
ASP, densest part	0.79	This study
Eastern South Dakota 1988	0.81	Dieter & McCabe, 1988
Wyoming	0.9	Collins, 1976
Massachusetts: Quabbin Reservation	1	Brooks, 1977
New Brunswick	1.09	Nordstrom, 1972
ASP, highest (lower Red House Bk.)	1.14	This study

ever reported is reached. Compared with the numbers of colonies per stream km in other studies, Allegany State Park contains both extremely low and extremely high colony densities (Table 1).

3.3. Intercolony Distances

Mean nearest neighbor distances of adjacent colonies—measured from lodge to lodge—at ASP averaged 1.11 km. Mean intercolony distances to neighbors upstream and downstream (including single colonies at certain streams) were 2.62 km at ASP (Houlihan, 1989) and 1.93 km in Fulton County (unpubl. data). The difference is not significant. Figure 3 summarizes these data. Other studies have reported mean nearest neighbor distances ranging from 0.51 km to 1.59 km. Some of these data are summarized in Table 2. They show that the mean nearest neighbor distances at ASP are relatively high.

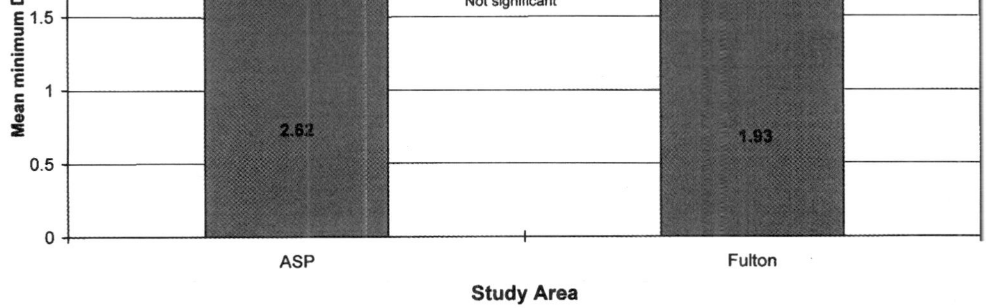

Figure 3. Intercolony distances at ASP and in Fulton County.

Table 2. Mean nearest neighbor distances (in km) between beaver colonies in several studies

Location	MNND	Reference
Newfoundland	0.51	Bergerud & Miller, 1979
Sierra Nevada	0.85 ± 1.2	Busher *et al.*, 1983
Massachusetts	0.93	Brooks *et al.*, 1980
ASP 1990	1.11 ± 0.97	This study; Schulte, 1993
Alaska	1.59	Boyce, 1974

3.4. Colony Size

The size of beaver families at ASP ranged from 4.1 to 6.12 in different years. For example, in 1987 mean colony size was 5.7 (Houlihan, 1989). Various other studies described family sizes ranging from 4.1 to 8.2 beavers per colony (Table 3). Thus, we observe at ASP the lower and medium family sizes described in other studies, but not the large family sizes, as reported from Colorado (Hay, 1958), Michigan (Shelton, 1979), Massachusetts (Brooks, Fleming and Kenelly, 1980), and California and Nevada (Busher, Warner and Jenkins, 1983) (Table 4).

3.5. Family Composition

Family composition at ASP was found to be typical for dense populations with a large percentage of families having an additional adult. In all cases where the identity of the supernumerary was known, it was a three-year-old offspring that stayed with its parents for another season. As many as 37.8% of the colonies contained an additional adult. This percentage lies in the middle of a number of other studies of dense beaver populations: Bergerud and Miller (1977) reported 22.2% of the colonies with a third adult, and Busher *et al.* (1983) 87.5% (Table 4). The four studies of dense populations, including

Table 3. Beaver family sizes in various studies

Location	X ± SD	Reference
ASP 1989	4.1 ± 2.38	This study
Alaska	4.1	Boyce, 1974
Montana	4.11	Easter-Pilcher, 1990
Newfoundland	4.2	Bergerud & Miller, 1972
HWF	4.28	This study
California: Sagehen Creek	4.8	Busher *et al.*, 1983
Michigan	5.1	Bradt, 1938
Wyoming	5.2	Collins, 1976
Newfoundland	5.3	Payne, 1982
ASP average 1989-1991	5.44 ± 2.45	This study
ASP 1987	5.7 ± 6.8	Houlihan, 1989
Ohio	5.9	Svendsen
ASP 1991	6.12 ± 2.6	This study
Colorado	6.3 ± 1.3[a]	Hay, 1958
Michigan: Isle Royale Natl. Park	6.4	Shelton, 1979
Massachusetts	8.1	Brooks *et al.*, 1980
Nevada: Little Valley	8.2	Busher *et al.*, 1983

[a][5.1 (willow) to 7.8 (aspen)]

Table 4. Family composition in high density beaver populations

Number of colonies	> 1 adult ♂	> 1 adult ♀	Percent of colonies with an additional adult	Reference
27	3	3	22.2	Bergerud & Miller, 1977
18	2	2	22.2	Brooks *et al.*, 1980
21	2.5[a]	2.5[a]	23.8	Payne, 1982
8	3	7	87.5[b]	Busher *et al.*, 1983
45	10	7	37.8	This study
119	20.5	21.5	35.8	Total
—	14.3	18.1	38.2	Average percent

[a]No sex given. In this table, the five colonies with supernumary "adults" are arbitrarily evenly divided between the male and female columns. Hence the values of 2.5.
[b]The 3 colonies with 2 "adult" males also had 2 "adult" females.

ours, report a mean of 34.9% of the families with supernumerary adults (which are most likely 3-year-old offspring). By way of a "metastudy", we found that of 323 colonies analyzed in these studies, 26.8 (see Tables 4 and 5 for an explanation of fractional values) had two adult males. Of the 119 colonies in high density populations, 20.5 (17.2%) had two adult males (most likely including 3-year-old individuals), while only six of the 204 colonies (2.9%) in low density populations had a second adult male ($x^2 = 20.4$; $P < 0.001$) (Table 5). Of the 119 high density populations, 21.5 had two adult females (18.1%), compared with 8.3 of 204 low density populations (4.1%) ($x^2 = 17.8$; $P < 0.001$).

3.6. Tail Scars

Increased population densities may result in more frequent competitive contacts between beavers. This may include intercolony agonistic encounters between dispersers and territory holders and between neighboring territory holders, but also intracolony aggression between parents and offspring, notably preceding dispersal to other sites. Most of these interactions occur during darkness and in vegetation cover, and cannot be easily observed and recorded. Therefore, the level of agonistic interactions has to be inferred from indirect evidence such as body scars. The most easily recorded signs of past injury are tail scars, mostly in the form of notches at the edge of the tail. Assuming that most of them result from fighting, and not from other injuries, such as from vegetation, by predators, or in traps, the number of tail scars can be used as an indicator of intraspecific strife.

At ASP 63.3 % of adult males had tail scars (females: 37%) (Figure 4). Males had more notches than females (Figure 5), but the difference is significant only in adults, and

Table 5. Family composition in low density beaver populations

Number of colonies	> 1 adult ♂	> 1 adult ♀	Percent of colonies with an additional adult	Reference
57	0	1	1.75	Bradt, 1938
10	0	0	0	Boyce, 1974
106	6.3[a]	6.3[a]	11.8	Novak, 1977
23	0	0	0	Svendsen, 1980
8	0	1	12.5	Busher *et al.*, 1983
204	6.3	8.3	4.9	Total
—	3.1	4.1	5.2	Average percent

[a]Novak estimated that 12.5 of 106 colonies had 3 beaver 2.5 years old or older. No sex given. In this table the figure of 12.5 is equally divided between columns for males and females. Hence, the value of 6.3.

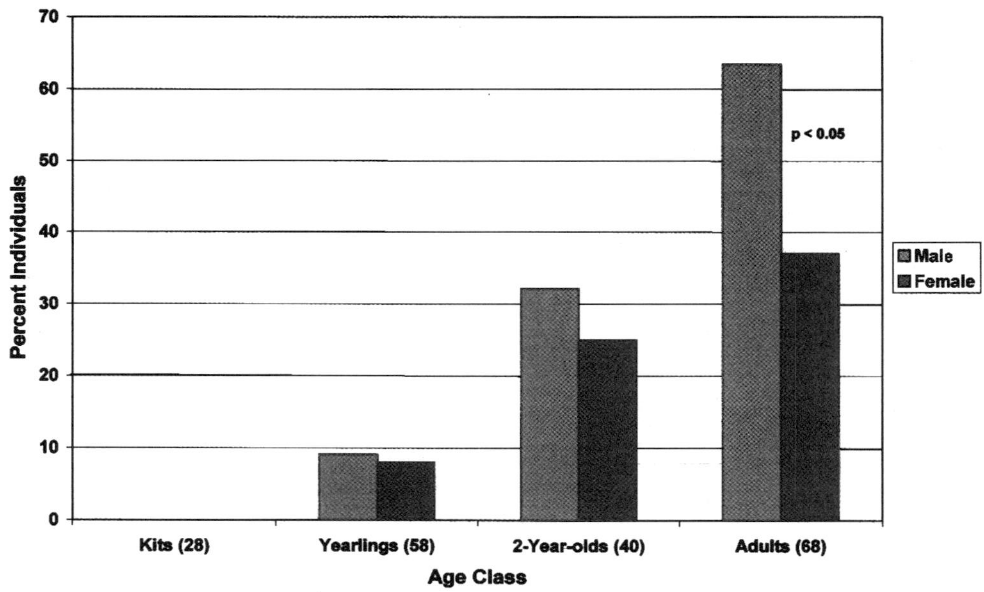

Figure 4. Percentage of beavers with tail notches at ASP, by age class.

not in 2-year-olds and yearlings. Kits never had notches, and adults had as many as 4 notches (entered in Figure 6 as category with > 2 notches). We did not find comparable published surveys of tail scars in exploited (harvested) populations.

3.7. Body Weights

Adult male beavers, sampled during the years 1985–92, weighed on average 17.95 kg (± 0.39 kg *SE*; $n = 43$) in spring, and 19.83 kg (± 0.87 kg *SE*; $n = 18$) in fall. Adult fe-

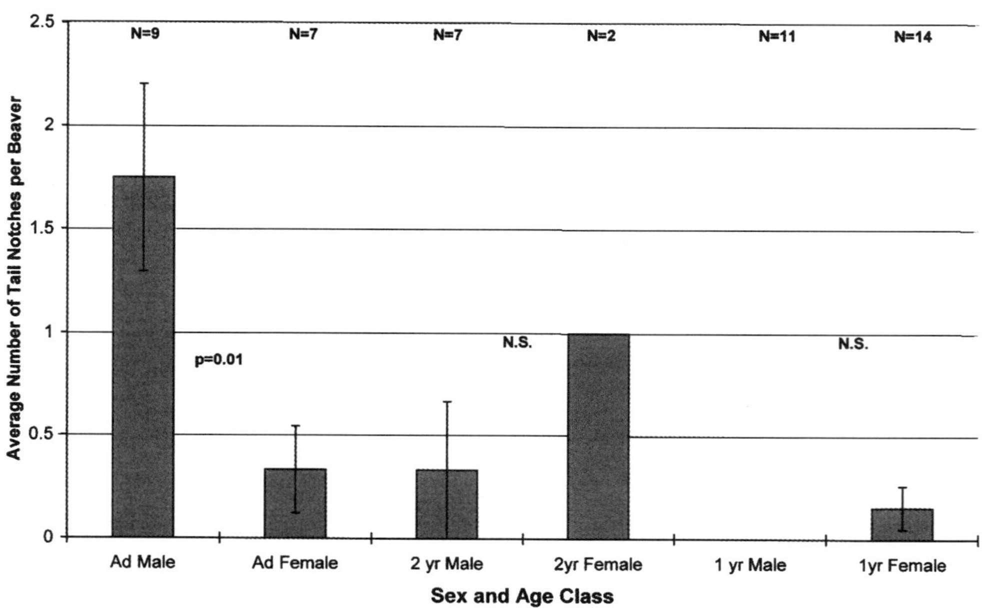

Figure 5. Average number of tail notches at ASP, by sex and age class.

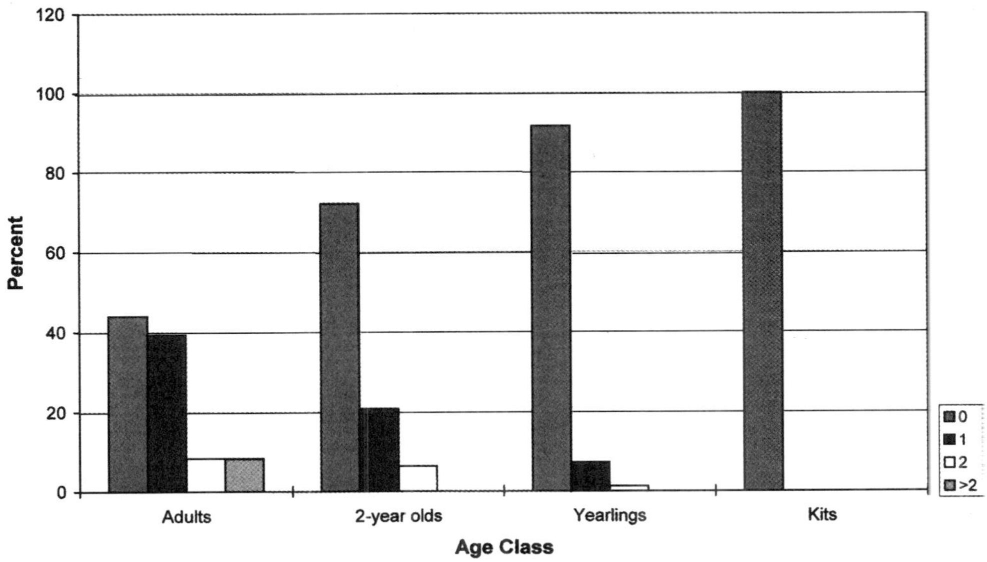

Figure 6. Percentage of beavers with zero, one, two, or more than two tail notches at ASP, by age class.

males weighed 18.62 kg (± 0.48 kg *SE*; *n*=38) in spring, and 18 8 kg ± (0.63 kg *SE*; *n* = 14) in fall (Figure 7). The overall mean weight of both sexes in both seasons was 18.71 kg. The difference between the fall and spring weights for both sexes combined is significant (*P* = 0.0029; 2x2 factorial ANOVA). The body weights of 2-year-olds at ASP are given in Figure 8. Body weights of yearlings and kits are presented in Figure. 9 and Figure. 10, respectively.

Compared with other studies, body weights at ASP were higher than in Alabama (17.7 kg; Wilkinson 1962), but lower than in Wood Buffalo National Park (both sexes in winter 20.79 kg, and in fall 21.34 kg; Novakowski 1967), and in the Mackenzie Delta (18 to 23 kg; Aleksiuk and Cowan, 1969), both in Canada.

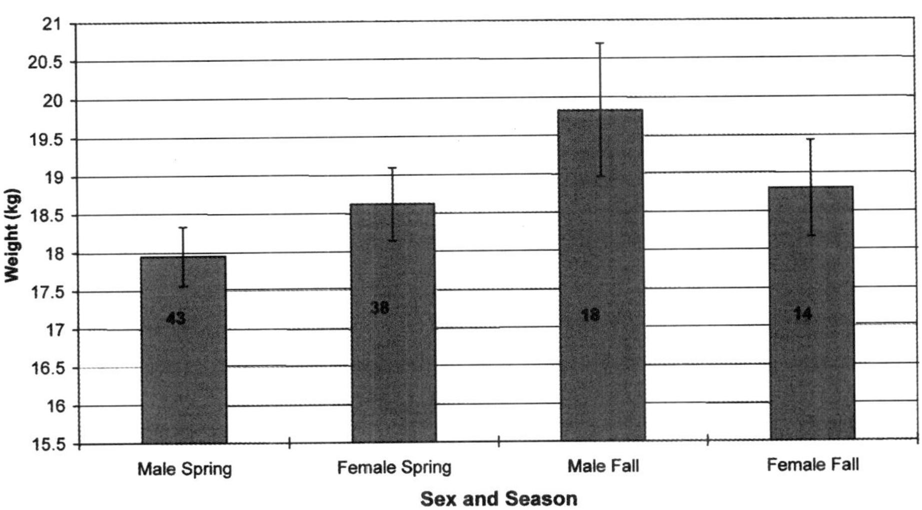

Figure 7. Body weights of adult beavers at ASP, in spring and fall.

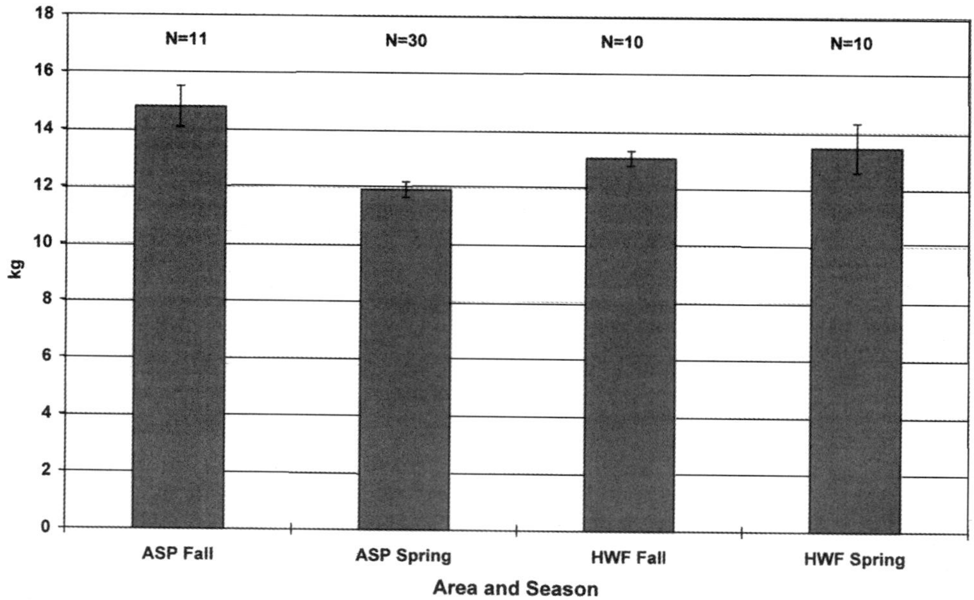

Figure 8. Body weights of 2-year-old beavers at ASP, in spring and fall.

As these studies cover vast latitudinal differences, the most meaningful comparison is with other populations in New York that differ from ASP mainly in being exploited, and therefore intentionally kept from becoming saturated. We compared the body weights of beaver at ASP with those at our other New York study area, Huntington Wildlife Forest (HWF) in the Adirondacks. In fall, the body weights of adult beaver of both sexes combined were greater at ASP than at HWF (HWF males: 17.31 ± 0.71 kg *SE*, $n = 13$; females: 17.78 ± 0.70 kg *SE*, $n = 12$). At HWF the overall mean body weight for both sexes and both seasons was 18.2 kg ($n = 53$), somewhat lower than at ASP (18.71 kg). The Hunt-

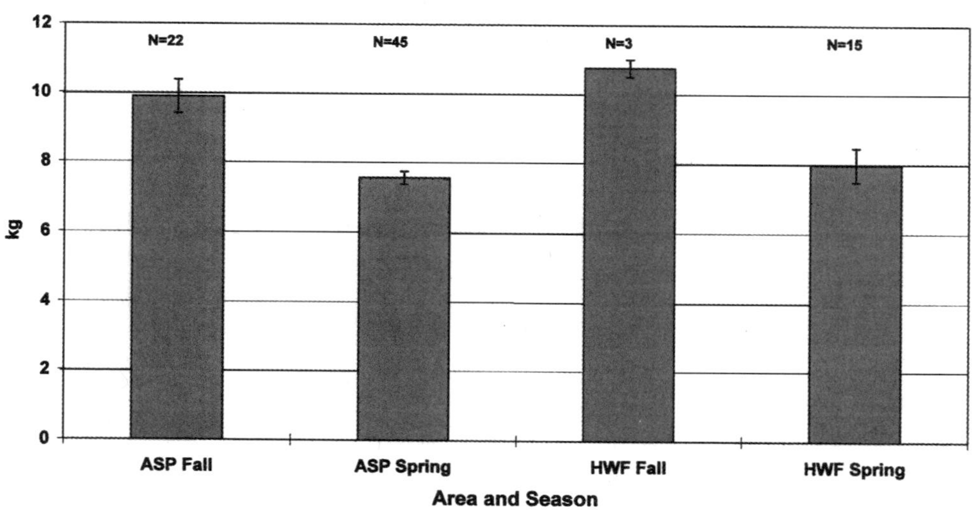

Figure 9. Body weights of yearling beavers at ASP, in spring and fall.

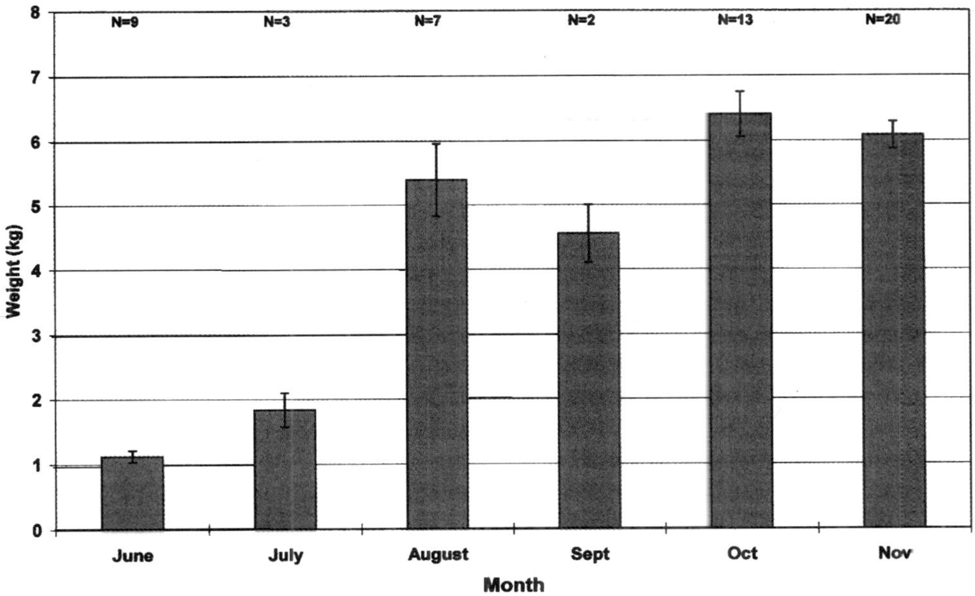

Figure 10. Body weights of beaver kits at ASP in the year they were born, by month.

ington Forest population itself is considered to be nutritionally challenged. Huntington Forest beavers live at higher elevations, and the region is more northerly than ASP. Because of this HWF beavers experience longer winters and have only a few species of trees available, none of them optimal for beaver. Beaver colonized the HWF area in large numbers early in the century in the wake of forest fires on 1500 ha that cleared the way for fresh growth of aspen, which now has been replaced by later successional trees (Sage,1983). At the time of our studies, the HWF population was growing from 16–18 colonies to 25–28 colonies as a result of reduced trapping in surrounding areas (Sage, pers. comm.).

Further data on body weights of winter-trapped Adirondack beaver (from the Moose River area) were obtained from the New York Department of Environmental Conservation (DEC). The body weights of these adults did not significantly differ from those of ASP adults in spring. Thirty-three adult males in the DEC sample averaged 17.76 kg (± 0.407 kg *SE*), and 43 males at ASP weighed 17.95 kg (± 0.39 kg *SE*). Sixteen adult females in the DEC sample averaged 18.76 kg (± 0.69 kg *SE*) ; and 38 adult females at ASP 18.52 kg ± (0.48 kg *SE*). Therefore, we found no significant decrease in body weights of adult beaver at ASP, compared with harvested populations.

3.8. Dispersal

In a saturated population, only marginal sites are available for colonization by 2- or 3- year-olds that have to leave their parental colony. Such marginal sites are small wet areas, usually along ditches or near culverts. Young beavers settling there, even only temporarily, are then often declared "nuisance beaver" and removed At what distances do successful dispersers finally settle? At ASP we collected information on 33 tagged beavers that had dispersed from 16 families living along nine streams. Nine were females, and 23 were males (the sex of one beaver was not noted). The mean dispersal distance for females was 8.876 km (± 3.630 km *SE*), and for males 4.013 km (± 1.238 km *SE*) (Sun and Müller-Schwarze, 1996). For comparison, 22 radio-equipped beaver in Montana moved as far as 7 km (mean maximum downstream movement in one of the two years (Van Deelen and Pletscher, 1996).

3.9. Food Choice

At ASP, beaver are very flexible in their foraging behavior in terms of tree species and size classes harvested. A comparison between the tree species utilized by the first beaver colonizing ASP in the late 1930s and those used by beaver there today can illuminate the possible removal of preferred species by the growing beaver population. The first beaver family to arrive at the Park in 1938 fed on seven tree species that are still available at many sites. These are, in descending order of preference: *Populus tremuloides* (quaking aspen), *P. grandidentata* (bigtooth aspen), *Salix* spp. (willow), *Amelanchier canadensis* (shadbush or juneberry), *Carpinus caroliniana* (musclewood or American hornbeam), *Prunus pennsylvanica* (pin cherry), and *Acer rubrum* (red maple; "only one tree") (Shadle and Austin, 1939). The beaver did not use "birches and maples". Unfortunately, we do not have information on the available tree species diversity around the first beaver sites at that time. The first six beaver colonies at the Park consumed 25 species of trees, belonging to 17 genera. Seven genera supplied 93% of the cuttings, while two genera - *Populus* and *Carpinus* - furnished 56.9% of the 5,424 cuttings counted (Shadle, Nauth, Gese and Austin, 1943).

We recorded the trees present and utilized at two beaver sites newly colonized in 1985. Thirteen species were utilized, while only Scotch pine (*Pinus sylvestris*) and white pine (*P. strobus*) were avoided. In addition to the species mentioned by Shadle *et al.* (1943) beaver consumed ash (*Fraxinus sp.*), hemlock (*Tsuga canadensis*), yellow birch (*Betula alleghaniensis*), black cherry (*Prunus serotina*), American beech (*Fagus grandifolia*), witch hazel (*Hamamelis virginiana*), sugar maple (*Acer saccharum*) and hawthorn (*Crataegus* spp.).

At continuously occupied sites, fewer tree species were available. At one site (Nr. 61) American hornbeam, American beech, black cherry, eastern hophornbeam ("ironwood", *Ostrya virginiana*) sugar maple, red maple, yellow birch, and hemlock were present. Of these, the first four were utilized. At a second site (Nr. 54B) only Red pine (*Pinus resinosa*), Scotch pine, black cherry, American hornbeam and downy serviceberry (*Amelanchier arborea*) were available. All of these five species were utilized. Finally at site Nr. 53, nine species were available, and six (beech, black cherry, sugar maple, willow, apple, and red pine) were utilized. The beaver did not utilize Scotch pine, Norway spruce, and red maple. Thus, continuously occupied sites suffered depletion of tree species, especially preferred ones. For the beavers that meant an impoverished selection of tree species, both in the sense of absence of preferred species and a smaller species ensemble from which to choose.

While the number of genera and species of trees available has not changed significantly, quantitative relationships may have changed. Shadle did not report what tree species were available (and in what ratios) at the time of his studies. But the beavers now cut trees in proportions different from those at Shadle's time. Figure 11 summarizes these changes for all six sites studied by Shadle in 1937–41 and all six sites (different ones) surveyed by us in 1984/85. Aspen and willow have decreased in the beavers' harvest, while black cherry, beech, witch hazel, ash, and pine have increased. We compared the percent cut of each tree species at each site then and now separately by t-test. Significant de-

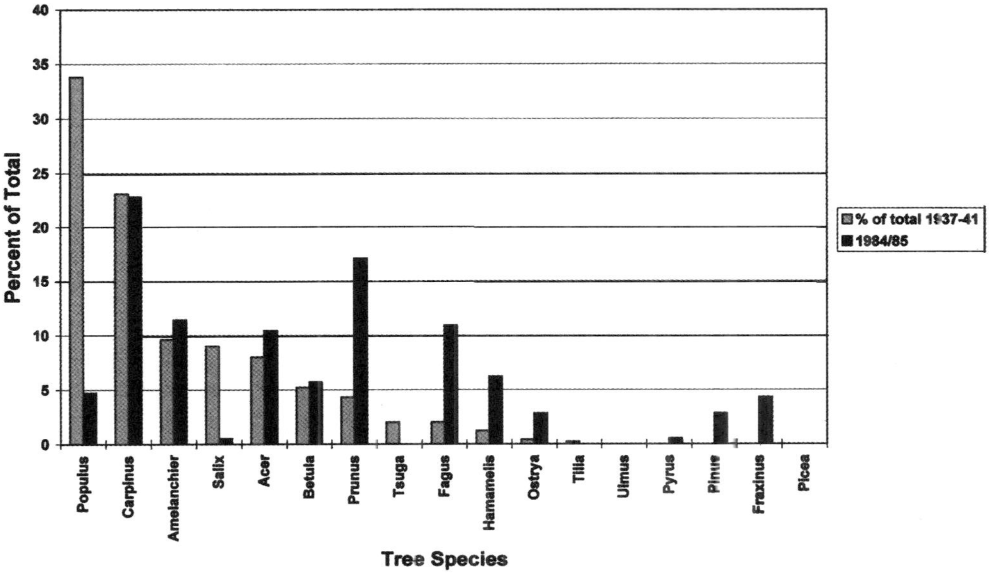

Figure 11. Distribution of trees cut by beavers at ASP in percent of the total number of trees cut. Light bars: trees cut by first beavers in 1937–41; dark bars: trees cut by long-established beavers in 1984/85.

creases in harvesting mean percentage for all ponds) were found for *Populus* (P = 0.007, one-tailed) and *Salix* (P = 0.025, one-tailed). A significant increase of cutting occurred in *Prunus* (P = 0.044, one-tailed), while the increases in cutting of *Acer* and *Fagus* were not statistically significant (Table 6). These findings indicate a depletion of the beavers' preferred species over the intervening decades.

3.10. Trail Lengths

The depletion of preferred tree species around beaver ponds led the beavers at some sites to travel greater distances to their preferred trees, (for example, aspen). At ASP, distances traveled to food sources fall into two distinct categories. Regular trails, a number of which are found at every pond, ranged from approx. 15 m to 80 m One sample of 18 trails at 6 sites averaged 31.5 m (*SD* ± 14.3 m; range 15.5 m to 72.9 m). On the other hand, at some colonies beavers used much longer trails if two conditions were met: the area around the pond was depleted of preferred trees, and an aspen stand existed at a greater distance from the pond. Six of these trails averaged 129 m (*SD* ± 31.5 m; range 65.3 m to 201 m) (Figure 12). In this bimodal distribution the means are significantly different from one another (t =3.677; P < 0.01). This indicates that the usual trails are short and distributed around the pond, while at other specific sites beavers maintain only one long trail to a distant aspen stand. These longer trails typically also traverse steep slopes.

3.11. Scent Marking and Responses to Pheromones

To test whether a dense beaver population scent marks more, we compared the average number of scent mounds per colony at ASP with that at HWF, and also with the numbers reported in the literature (Table 7). The average number of scent mounds per colony

Table 6. Statistical differences in the percentages of trees cut by beavers in 1937–41 (Shadle *et al.*, 1943) and 1985/85 (this study)

Species	Mean percent cut in 1937-41	Mean percent cut in 1984/85	Statistical difference (*P* one-; two-tailed)
Populus	44.1	2.8	0.007; 0.014
Carpinus	17.3	17.5	ns
Amelanchier	6.6	9.4	ns
Salix	10.6	1.3	0.025; 0.05
Acer	7.3	15.8	ns
Betula	4.8	4.3	ns
Prunus	4.3	11.2	0.044; 0.087
Tsuga	1.3	0	ns
Fagus	3.1	13	ns
Hamamelis	1.6	4.3	ns
Ostrya	0.4	2.8	ns
Tilia	0.2	0	ns
Ulmus	0.1	0	ns
Pyrus	0.1	1.3	ns
Pinus	0.1	4.0	ns
Fraxinus	0.1	3.1	ns
Crataegus	0	0.4	ns

at ASP is similar to the number at HWF. However, the average number of mounds per colony at ASP is much lower than the average numbers reported at the Quabbin Reservation, Massachusetts (Bollinger, 1980; Hodgdon, 1978).

Overall means may obscure intra-population differences in scent marking, according to local population density. The number of scent mounds at a beaver site is positively correlated with the number of other colonies within 5 km upstream or downstream. The correlation between number of beaver colonies within 5 km upstream or downstream and the number of scent mounds is $r^2 = 0.665$; $P = 0.0001$, $n = 20$ (Houlihan, 1989). The correlation between the number of scent mounds at a colony and the distance to the nearest neighbor downstream is $r^2 = 0.385$; $P = 0.01$, $n = 20$; and to the nearest neighbor upstream $r^2 = 0.387$; $P = 0.03$, $n = 20$ (Houlihan, 1989). More recently, the same relationships have been found in the Eurasian beaver, *Castor fiber*, in the Netherlands (Rosell and Nolet, 1997). These authors found a correlation between numbers of scent marks and numbers of neighboring colonies to be $r = 0.753$; ($P = 0.003$, $n = 13$); and between number of scent marks and mean distance to all other beaver territories $r = -0.710$ ($P = 0.007$, $n = 13$). I f r^2 values are calculated for these reported r values they are $r^2 = 0.567$ and $r^2 = 0.504$, respectively.

Responses to experimentally presented scent marks were stronger at ASP than in two other New York populations (HWF and Cranberry Lake Biological Station), possibly indicating heightened territorial competition for sites. Number of land visits to experimental scent mounds was 58% at ASP, and 25% at HWF. This is the only significant difference between the populations; all other measures of responses to pheromones were higher at ASP, but not significantly so. Although beavers have been exposed to experimental pheromonal stimuli (castoreum and anal gland secretion) in other areas, the investigators have measured the responses in different ways, so that the results of the different studies cannot be compared easily. In our studies, durations of sniffing the artificial scent mounds were averaged 22.3 seconds at ASP (Houlihan 1989), and 22.2 seconds at HWF (unpublished). At the unexploited population at Quabbin Reservoir in Massachusetts adult male beaver

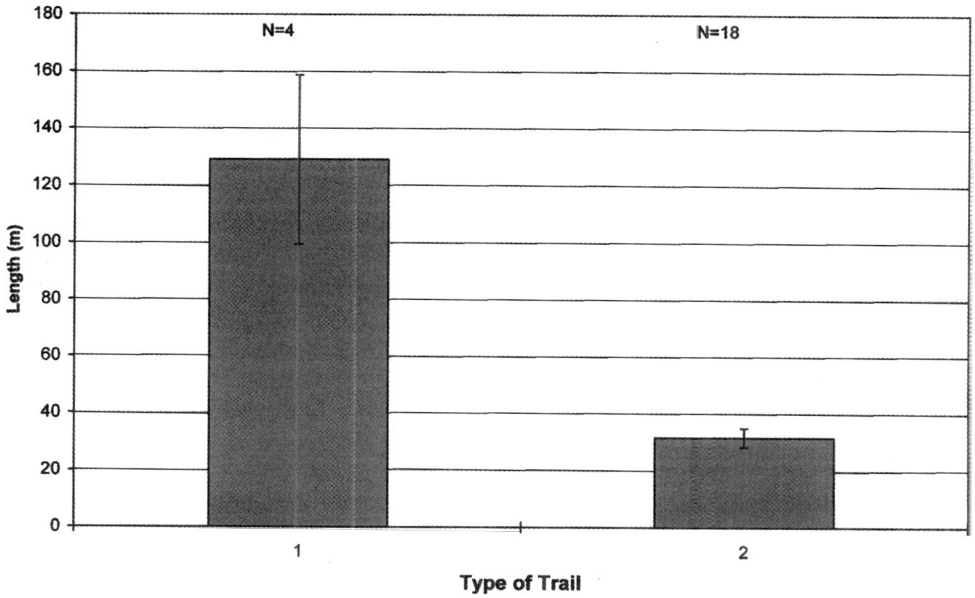

Figure 12. Beaver foraging trails of two different lengths at ASP.

sniffed artificial male castoreum marks for 19.3 ± 6.0 seconds, and female castoreum marks for 10.0 ± 0.6 seconds. The corresponding times for adult females were 13.2 ± 2.5 seconds, and 11.6 ± 3.7 seconds, respectively (Hodgdon, 1978). This is lower than at ASP.

4. DISCUSSION AND CONCLUSIONS

Our data indicate that unexploited, saturated populations differ in some ways from exploited, continuously thinned out populations:

1. Beavers colonize steeper stream gradients.
2. Families contain a large percentage of 3-year-olds (young that stay one more year).

A number of studies confirmed that beavers adjust their age of first breeding to the population density. In unexploited (high density) populations many beaver delay maturity beyond 2 years (and are presumed to stay at their natal colony). In heavily exploited (low

Table 7. Average number of scent mounds per colony in different beaver populations

Beaver population	Number of colonies	Mean # scent mounds (± SE)	Range	Reference
ASP	20	38.5 ± 8.5	0–133	Houlihan, 1989
HWF	6	28.7 ± 8.4	10–67	This study
Quabbin Reservoir, Massachusetts 1971-73	8	87.9 ± 11.7	12–125	Hodgdon, 1978
Quabbin Reservoir 1977	3	164.3 ± 50.9	67–239	Bollinger, 1980
Quabbin Reservoir 1978	3	108.0 ± 34.0	40–144	Bollinger 1980

density) populations, many beavers breed as yearlings (Payne, 1982; Dieter, 1992). In New York, yearlings started to breed when only 40% of potential sites were occupied by beavers (Parsons and Brown, 1979). Unexploited populations seem to have longer generation times because of the delay in dispersal compared to exploited populations. In the latter, 2-year-olds, 3-year-olds and even yearlings commonly reproduce. The trade-off seems to be when to reproduce and not the number of offspring produced.

3. Preferred tree species become depleted and beavers feed on less palatable species.
4. Beavers extend their trails to distant foraging areas.

Similarities with exploited populations include:

1. Densities of colonies per stream kilometer did not differ from those reported in other studies. In growing populations, the series of beaver colonies along a stream cannot be compressed beyond a certain point. Usually, the minimum distance between colonies lies around 0.9 km. In Boyce's (1981) study, inter-colony distances below 1 km were rare; the lowest distance measured was 0.48 km., and inter-colony distances did not decrease.
2. Body weights did not differ from those in other studies.
3. Family size is within the range of other populations.
4. The numbers of scent mounds in territories are not reduced or increased.
5. The response to artificial scent marks does not differ in intensity from other areas.

These findings provide some "end points" of what to expect from growing beaver populations. Wildlife managers may find some characteristics useful as diagnostic tools to assess the status of beaver populations. Managers often try to intervene in populations deemed "too dense", yet beavers can maintain intercolony distances, family size, and body weight, despite high population densities. Managerial decisions then address conflicts between beavers and humans, and not necessarily any requirements of the beaver themselves.

ACKNOWLEDGMENTS

Over the many years of these studies many graduate and undergraduate students have helped gathering data in the field. We especially thank Dr. Lixing Sun and Tim Schwender for collecting data and discussing them with us. We thank the managers and staff of Allegany State Park and Huntington Wildlife Forest for logistical and other support. The work was partially supported by an NSF grant to DMS, and a grant from the Theodore Roosevelt Fund to BAS.

REFERENCES

Aleksiuk, M., and Cowan, I. McT., 1969, Aspects of seasonal energy expenditure in the beaver (*Castor canadensis* Kuhl) at the northern limit of its distribution. *Canad. J. Zool.* **47:** 471–481.

Bergerud, A. T., and Miller, D. R., 1977, Population dynamics of Newfoundland beaver. *Canad. J. Zool.* **55:** 1480–1492.

Bollinger, K. S., 1980, Scent marking behavior of beaver (*Castor canadensis*) M.S. thesis, Univ. Massachusetts, February 1980, 186 pp.

Boyce, M. S., 1974, Beaver population ecology in interior Alaska. M.S. thesis, Univ Alaska (Fairbanks); 161 pp.

Boyce, M. S., 1981, Beaver life history response to exploitation. *J. Appl. Ecol.* **18:**749–753.

Bradt, G. W., 1938, A study of beaver colonies in Michigan. *J. Mammal.* **19:**139–162.

Brooks, R. P., Fleming, M.W., and Kennelly, J. J., 1980, Beaver colony responses to fertility control: evaluating a concept. *J. Wildl. Manage.* **44:** 568–575.

Busher, P. E., 1987, Population parameters and family composition of beaver in California. *J. Mammal.* **68:** 860–864.

Busher, P. E., Warner, R. J., and Jenkins, S. H., 1983, Population density, colony composition, and local movements in two Sierra Nevada beaver populations. *J. Mammal.* **64:** 314–318.

Collins, T. C., 1976, Population characteristics and habitat relationships of beavers, *Castor canadensis*, in North-west Wyoming. Ph.D. Dissertation, Univ. Wyoming, Laramie, Wyoming.

Dieter, C. D., 1992, Population characteristics of beavers in eastern South Dakota. *Am. Midl. Nat.* **128:** 191–196.

Dieter, C. D. and McCabe, T. R., 1988, Beaver crop depredation in eastern South Dakota. *Prairie Natur.* **20:** 143–146.

Easter-Pilcher, A., 1990, Cache size as an index to beaver colony size in Northwestern Montana. *Wildl. Soc. Bull.* **18:** 110–113.

Hay, K., 1958, Beaver census methods in the Rocky Mountain region. *J. Wildl. Manage.* **22:** 395–402.

Hodgdon, H. E., 1978, Social dynamics and behavior within an unexploited beaver (*Castor canadensis*) population. Doctoral Dissertation, Univ. Massachusetts, August 1978, 292 pp.

Houlihan, P. W., 1989, Scent mounding by beaver (*Castor canadensis*): Functional and semiochemical aspects. M.S. Thesis. S.U.N.Y. College of Environmental Science and Forestry, Syracuse, N.Y., USA. 184 pp.

Nordstrom, W. R., 1972, Comparison of trapped and untrapped beaver populations in New Brunswick. M S. Thesis, Univ. New Brunswick, Fredericton, 104 pp.

Novak, M., 1977, Determining the average size and composition of beaver families. *J. Wildl. Manage.* **41:** 751–754.

Novakowski, N. S., 1965, Population dynamics of a beaver population in northern latitudes. Ph. D. Diss., Univ. Saskatchewan (Saskatoon); 154 pp.

Novakowski, N. S., 1967, The winter bioenergetics of a beaver population in northern latitudes. *Can. J. Zool.* **45:** 1107–1118.

Parsons, G. R., and Brown M. K., 1979, Yearling reproduction in beaver as related to population density in a portion of New York. Trans. Northeast Sect., *The Wildl. Soc., Fish and Wildl. Conf.* **36:** 188–191.

Payne, N. F., 1982, Colony size, age, and sex structure of Newfoundland beaver. *J. Wildl. Manage.* **46:** 655–661.

Rosell, F., and Nolet B. A., 1997, Factors affecting scent-marking behavior in Eurasian beaver (*Castor fiber*). J. *Chem. Ecol.* **23:** 673–689.

Sage, R. W. jr., 1983, An evaluation of several herbicide applications used for site preparation in northern hardwood stands. MS thesis SUNY College of Environmental Science and Forestry, 160 pp.

Schulte, B. A., 1993, Chemical communication and ecology of the North American beaver (*Castor canadensis*). Ph.D. Dissertation, S.U.N.Y. College of Environmental Science and Forestry, Syracuse, N.Y., USA, 194 pp.

Shadle, A. R., and Austin T. S., 1939, Fifteen months of beaver work at Allegany State Park. *J. Mammal.* **20:** 299–303.

Shadle, A. R., Nauth A. M., Gese E. C., and Austin T. S., 1943, Comparison of tree cuttings of six beaver colonies in Allegany State Park, New York. *J. Mammal.* **24:** 32–39.

Shelton, P. C., 1979, Population studies of beavers in Isle Royal National Park, Michigan, in: *Proc. First Conf. Sci. Res. in the National Parks*, Volume 1 (R. M. Linn, ed.), pp. 353–356.

Sun, L. and Müller-Schwarze D., 1996, Statistical resampling methods in biology: A case study of beaver dispersal patterns. *Amer. Journal. of Mathematical and Management Sciences* **16:** 463–502.

Svendsen, G. E., 1980, Population parameters and colony composition of beaver (*Castor canadensis*) in southeast Ohio. *Am. Midl. Natur.* **104:** 47–56.

Van Deelen, T. R., and Pletscher D. H., 1996, Dispersal characteristics of two-year-old beavers, *Castor canadensis*, in western Montana. *Canad. Field Naturalist* **110:** 318–321.

Welsh, R. G., and Müller-Schwarze D., 1989, Experimental habitat scenting inhibits colonization by beaver, *Castor canadensis. J. Chem. Ecol.* **15:** 887–893.

Wilkinson, P. M., 1962, A life history study of the beaver in east-central Alabama. M.S. thesis, Auburn Univ., 91 pp.

INDEX